John Cohn Hemmeter

On the Comparative, Physiological Effects of Certain Members of the Ethylic Alcohol Series

CH4O to C5H12O

John Cohn Hemmeter

On the Comparative, Physiological Effects of Certain Members of the Ethylic Alcohol Series
CH4O to C5H12O

ISBN/EAN: 9783744669726

Printed in Europe, USA, Canada, Australia, Japan

Cover: Foto ©berggeist007 / pixelio.de

More available books at **www.hansebooks.com**

ON THE

COMPARATIVE PHYSIOLOGICAL EF-

FECTS OF CERTAIN MEMBERS OF THE ETHYLIC

ALCOHOL SERIES (CH_4O to $C_5H_{12}O$) ON THE ISOLATED

MAMMALIAN HEART.

-- By --

John C. Hemmeter M.D. Candid. Philos.

Johns Hopkins University.

L i t e r a t u r e.
----

1. Liebig's Annalen (d. Chemie u. Pharmacie LXXIII p. 213) for 1850 by T. Schlossberger.

2. "Handbuch der Arzneimittellehre (1880 p. 340)" Nothnagel u. Rossbach.

3. A. F. A. Gros "L'action de l'Alcool amylique sur l'Organisme."

4. Comptrend LXXXI p. 152-154 and LXXXIII p. 80. Dujardin Beaumetz and Audigé.

5. Journal of the Chemical Society, Vol. 29, 1886 and Vol. 30, p. 539.

6. Bulletin gen. de therap. Sept. 30, '84, p. 251. Dujardin Beaumetz and Audigé.

7. Stenberg, (Nordiskt Med. Arkiv. B X No. 21.)

8. J. Dogiel, (Phlüger's Archiv. for 1874 Vol. VIII (605.)

9. J. Dogiel, in Woods' Handbook of Therapeutics.

10. Dr. H. Zimmerberg, Untersuchungen über den Einfluss des Alcohols auf die Thätigkeit des Herzens. Dorpat 1869.

11. Woods' Therapeutics.

12. Dr. J. W. Warren (Alcohol Again, Boston Medical and Surgical Journal, July 7th and 14th, 1887).

13. Dr. Edward Smith, (On the action of foods on the respiration during the primary processes of digestion. Phil. Trans. Vol.

149, p. 731.)

14. Sydney Ringer M.D. and Harrington Sainsbury M.D. London Practitioner Vol. 30, 1883, p. 539. (Observations on the Relative Effects of certain Members of the Ethylic Alcohol Series on the Ventricle of the Frog's heart.)

15. Specific Gravity Tables of Landolt and Börnstein.

16. Bergeron & L'Hôte (Compt. rend. CXI No. 7) also Gaz. heb'd med N. 35 p. 568.

17. Vol. III p. 483 of Studies from the Biological Laboratory of the Johns Hopkins University, Dr. J. R. Duggan on the "Influence of Alcohols on the conversion of Starch by Diastase".

18. Dr. F. S. Lee in Studies from the Biol. Lab. of the Johns Hopkins University Vol. III.

19. Buchheim (Archiv. der Heilkunde 1866, p. 124)

20. Austie (Final Experiments etc.)

21. Schmiedeberg (Petersburg Medic. Zeitschrift 1868, B. 14 S. 92; also Virchow's Archiv. Vol. 51 p. 171.

22. Herrman (Archiv. f Anat. u. Physiolog. von Reichert u. Dubois-Reymond 1866 S 27.)

23. Maroaud (L'alcool son action physiologique etc. Paris 1872, p. 14).

24. Jaillet & Hayem (Virchow & Hirsch--Jahresbücher d. Gesammt. Medizin 1884, p. 369).

25. C. N. Schulz (Wirkung des Branntweins in der Trunksucht

Hufeland's Journal fur prakt. Heilkunde, 1841, April).

26. Lallemand Perrin & Duroy (Du role de l'alcool etc.etc. Paris, 1860).

27. Baer (Alcoholismus).

28. Vol. II p. 313, plate XV, this Journal.

29. Proceedings of the Medico-Chirurgical Faculty of Maryland (April 29, 1882).

30. Prof. H. N. Martin and L. T. Stevens. Studies fr. the Biol. Lab. of the Johns Hopkins University, Vol. 4, p. 213

31. C. Binz Vorlesungen über Pharmakologie.

32. J. Beachamp Compt. Rend T. LXXXIX p. 573
    Sur la presence de l'alcool dans les tissus animaux pendant la vie et apres la mort.

33. Bechamp Journ. Chem. Soc. Vol. 36 p. 763

34. " Archiv. de Physiol. Normal et Pathol.

35. A. " " " " 1882, p. 28

36. " " Compt. rend. LXXXI p. 236-239

37. Maurice Perrin (Sur l'action physiologique de l'alcool) Bull de'l Acad. de Med. N. 17 p. 521

38. Jaillet. De l'alcool son combustion, son action physiologique, son antidote, Thesis. Paris.

39. N. S. Davis, Trans. American Med. Assoc. 1885, p. 577

40. Brown Seynard, Journ. de la Physiolog. 1859 p. 467

41. P. Ruge, Virchow's Archiv. Bd XLIX, . 365.

42. C. Binz, Virchow's Archiv. B LI p. 153.

43. London Practitioner Vol. III 1869, Vol. V, 1870, Vol.XXX.

44. Ringer & Fichards, Lond. Lancet 1866, p. 208

45. Journal of Anat. & Physiol. Vol. VIII 1874, p. 232

46. Parkes & Wollowicz. Trans. Roy. Soc. 1870.

47. Anstie Stimulants and Narcotics.

48. Brunton L. T. Practitioner 1876 p. 57 and p. 118.

Synopsis of the Thesis.

I.	History and literature of the subject	p.	1
II.	Chemistry of the alcohols	"	16
III.	Influence of Alcohol on the constituents of the blood.	"	18
IV.	The method	"	23
V.	Results. Explanation of Tables	"	27
VI.	Relation between weight of animals and effect of alcohols	"	38
VII.	Position of Ethyl Alcohol in the Series	"	44
VIII.	Causes of the diminution of the work of the isolated heart under alcohol.	"	47
IX	Biographical note	"	51

The literature on the subject of the physiological activity of alcohols is very extensive; its abnormal growth appears to those who study and make an effort to keep abreast of the progress and advancement of experimental therapeutics, out of all proportion to any real increase in our knowledge of the subject. In attempting therefore, a very brief review of the scientific work that has been done in this direction, all speculative literature is excluded.

One of the earliest series of investigations on the comparative physiological effects of alcohols is reported in Liebig's Annalen (d. Chemie u. Pharmacie LXXIII p. 213) for 1850 by T. Schlossberger who worked under Griesinger. This report is very incomplete, stating simply that chemically pure methyl and amyl alcohol was experimented with and that their action is similar to that of Ethyl alcohol. As regards intensity of action amyl alcohol does not exceed Ethyl and methyl alcohol. The animals experimented upon were cats, dogs and rabbits. The statements of method employed and of the results is quite unsatisfactory.

Nothnagel and Rossbach in their Handbuchder Arzneimittellehre" (1880, p. 340) give an account of experiments with the first five members of the ethylic alcohol series. The action of

all are said to be similar in quality but different in quantity. The poisonous property increases with the molecular structure being least marked with methylic and most marked with amylic alcohol. On the authority of a thesis by one A. F A. Gros "L'action de l' Alcool amylique sur l'Organisme" Strassbourg 1883, Nothnagel and Rossbach give a definite numerical relationship stating Amylic to be thirty times as strong as methylic and fifteen times as strong as ethylic alcohol. From this one would be justified in concluding that "Gros" found ethyl to be fifteen times stronger than methyl alcohol.

Dujardin Beaumetz and Audige have published a series of investigations on the comparative toxic properties of the alcohols in Compt rend LXXXI p. 152-154 and LXXXIII p. 80 of which we find abstracts in the Journal of the Chemical Society. Vol. 29, 1886, Vol. 30, p. 539. These experiments consisted in administering varying proportions of ethylic propylic, butylic and amylic alcohols to various dogs, the quantities in some cases being passed into the stomach, in others injected under the skin. The weight of the doses administered was in every case kept in ratio to the weight of the animal operated upon. The following conclusions are drawn from the results.

I. The toxic properties of fermentation alcohols follow their atomic composition in a sort of mathematical order.

The higher the figures representing the atomic composition the more considerable is the toxic effect. This is equally true whether they are injected under the skin or introduced into the stomach.

II. For the same alcohol the toxic action is greater when the dose is administered through the stomach than subcutaneously, in the latter case the toxic effect is increased by dilution.

III. The toxic phenomena are the same for all alcohols save in the degree of intensity. The injuries caused follow in a progressive scale, from the ethylic to the amylic alcohol. The injuries to the mucous membrane of the stomach being just as great whether the alcohols are administered subcutaneously or are ingested by the stomach. Severe congestion of the small intestines was noticed in some cases, in whichever way the alcohols had been administered and moreover it was noticed that for the same alcohol, the congestion and pulmonary apoplexy were most frequent when the alcohol had been administered by the stomach.

In the "Bulletin gen. de therap" Sept. 30th, 1884 p. 251, Dujardin Beaumetz and Audige give the following table of toxic doses of alcohols per Kilogram weight of animal causing death in 24 hours.

Table of Dujardin Beaumetz and Audige of toxic doses of alcohols in Grams.

	Pure	Diluted
Ethyl	8	7.75
Propyl	2.90	3.75
Butyl	2	1.85
Amyl	1.70	1.30--1.80
Methyl		7.

Dujardin, Beaumetz and Audige's experiments have been in part repeated by Stenberg (Sten Nagra experimentela bidrag till belysande of frågan om det inflytaupe, som bräuvinets förorenin gar hafrapa dess fysiologis ka verkningar) Nardiskt Med. Arkiv BX. N. 21.

This observer believes that the methods of the French investigators have led them to defective results. By injecting the alcohols subcutaneously in the doses used by them he found them to cause great pain, irritation and suppuration. He holds that the amount of Glycerine with which Dujardin Beaumetz and Audige have in part injected their alcohols subcutaneously, is itself sufficient to cause death.

In Phlüger's Archiv for 1874 Vol. VIII (605) are published a series of resumes and experiments by J. Dogiel, which are often referred to as proving the alcohol really stimulates the heart

and raises blood pressure. According to Wood's Handbook of Therapeutics, it is affirmed by Dogiel, that arterial pressure is at first increased and then diminished. During the latter state the vaso motor centres are found to be incapable of responding to stimulation. The rate of the heart beat is stated to be at first increased, then diminished, then increased; the first increase being owing to the stimulation of the accelerators, the diminution to stimulation of the vagus and the final increase to paralysis of the same. This experimental evidence to H. C. Wood Jr. seems to confirm the clinically known fact, that alcohol in moderate amounts powerfully stimulates the heart and circulation, but in poisonous doses diminishes both the force and frequency of the pulse.

Another contribution to our knowledge of the physiological action of alcohol, which has scientific merit is that of Dr. H. Zimmerberg (Untersuchungen über den Einfluss des Alcohols auf die Thatigkeit des Herzens.) (Dorpat 1869). These studies were carried on with men and frogs, rabbits and cats. Zimmerberg asserts that alcohol causes a decided reduction of the pulse rate and of arterial pressure. If the vagi were cut during the depression, a sudden rise beyond the normal point in both pulse rate and arterial pressure occurred. When the alcohol was injected towards the heart into the jugular vein, the fall in the arterial pressure was almost instantaneous and very marked. After section of the vagi the alcohol lessened the arterial pressure without affecting

the pulse rate (so far quoted from Wood's Therapeutics). As these observations are opposed to clinical experience it was presumed that some fallacy must be underlying them, which was by H. C. Wood Jr. found in the enormous doses of alcohol used by Zimmerberg.

According to Dr. J. W. Warren ("Alcohol Again, Boston Medical and Surgical Journal, July 7th and 14th, 1887), the dissertation of Zimmerberg does show large doses for frogs, rabbits, and cats, but it cannot be urged that those given to men were unduly large.

On the whole it seems to us that in some of his conclusions, Zimmerberg has the weight of later evidence in his favor.

It will not prove expedient to quote the experiments of Parkes and Wollowicz upon man; here, to any extent, they show the accelerating effect upon the heart beat by sphygmographic tracings, which however give no distinct indication of increased arterial pressure.

These results are similar to those of Dr. Edward Smith. (On the action of foods on the respiration during the primary processes of digestion. Phil. Trans. Vol. 149, p. 731). The experiments of Parkes and Wollowicz were not carried out with pure alcohol but with brandy, and the results of both of these observers together with those of Dr. Edward Smith, have been superseded by

the more accurate experiments of Prof. H. N. Martin and L. T. Stevens B. A. of this University published in a previous number of this Journal. It is not intended to recapitulate the investigations of Prof. H. N. Martin here, they have become too well known since their publication and the present series of experiments are a continuation or rather an outgrowth from them.

We come now to speak of a series of experiments by Sydney Ringer M.D. and Harrington Sainsbury M.D. published in Vol. 30, 1883, p. 339 of the London Practitioner. (Observations on the Relative Effects of certain Members of the Ethylic Alcohol Series on the Ventricle of the Frog's heart.)

"The object of these experiments was to determine the relative physiological activity of certain members of the above series. The dose of any particular alcohol requisite to abolish the function of the frogs ventricle, gave the measure of the activity of this alcohol. The following alcohols were taken: methylic, ethylic, propylic, butylic, amylic--the normal alcohols of the first three were selected; of the last two the isomerics were taken, viz: iso and pseudo - butyl alcohol and iso amyl alcohol. In addition to these, isopropylic alcohol was experimented with. The first three members alone of the series, are consequently more strictly comparable if the object in view be the relation between molecular constitution and physiological activity. But from a practical point of view the whole series must be brought in, for

of the above, propylic, butylic and amylic alcohols, constitute for the most part the fusel oil which contaminates beer, wine and spirits.

"The experiments of Ringer and Sainsbury were made with a Roy tonometer; the circulating fluid consisted of a solution of dessicated bullocks blood in water of about the concentration of normal blood; this was further diluted with two and a half times its volume of saline, 0.6 per cent. The ligature was applied as nearly as possible in the auriculo-ventricular groove. The heart thus fed with the blood mixture as a rule began soon to beat spontaneously. At definite intervals of time the drug was now added to the whole mass of the circulating fluid. The dosage was maintained uniform and was such as previous experiment had shown to be sufficient to arrest the heart within an hour approximately. The restriction of time was for the purpose of eliminating the error due to the natural process of dying.

The functions of the ventricle were in each case completely abolished by the end of the experiment, so that the ventricle neither beat spontaneously nor responded to electric excitation. This is all that Ringer and Sainsbury state of their method, the account is not as explicit and accurate as might be desired. They state that all of the alcohols examined arrested the heart in diastole. The following quantities (in Minims) of the various alco-

hols were required to accomplish this effect:

Methyl	Ethyl	Propyl	Butyl	Amyl
CH_4O	C_2H_6O	C_3H_8O	$C_4H_{10}O$	$C_5H_{12}O$
205.5	114	59.3	17	6.6

The same quantities expressed in cubic centimetres would appear as follows:

Methyl	Ethyl	Propyl	Butyl	Amyl
13.1 c.c.	6.73 c.c.	3.5 c.c.	1 c.c.	0.4 c.c.

Attention is called to the rapidly ascending ratio of activity, as we pass from the lower and simpler alcohol to the higher and more complex members of the series, hence the first proposition is: that activity increases with the complexity.

Secondly, our attention is called to the fact that the numbers of the first three members stand to each other very nearly in geometric progression. The ethylic number is half the methylic the propylic half the ethylic. The butylic and amylic numbers do not fit in the series it is true, they do not however show any falling off in activity, but rather an increase. Continuing Ringer and Sainsbury explain their physiological results, by stating that a difference of CH_2 which is that by which each molecule in the series differs from its preceding and succeeding terms, is capable of halving or doubling, as the case may be, the activity of the molecule, each CH_2 group may be said in a way to have its physiological equivalent; the proposition deduced in short is, that as

long as the arrangement of the units are similar, as long as the lines of the structure remain the same, the addition or subtraction of each unit will carry with it a definite and constant difference. A certain Mr. Perkin is said to have demonstrated as much for certain of the alcohol series, having shown that under certain definite conditions the addition of each CH_2 corresponds to a definite quantitative difference in the degree of polarization of light which the particular alcohol is capable of effecting. Three main conclusions are finally arrived at, viz:

    I.   The qualitative similarity of action of the different members of the alcoholic series.

    II.   The general quantitative relationship, viz: that as the complexity of the molecule increases the physiological activity increases.

    III.   The probability of a further quantitative relationship, viz: that the constant chemical difference is corresponded to by a constant physiological difference, that each CH_2 group increases the activity by a definite amount.

    The work of Ringer and Sainsbury gives no evidence of any primary stimulation or increased frequence or force of the heart's contractions --on which point Dogiel attempted to confirm clinical experience, the results of the authors above, rather confirm those of Prof. Martin, that the effect described is not the result of direct action on the cardiac tissues.

The objections that can be urged against the work of Ringer and Sainsbury as reported in the Practitioner for 1883 are first, lack of clearness in the statement of the method, secondly, employment of unusually strong solutions of the alcohols for frogs. It is difficult to understand how the results stated in this paper were ever obtained after such strong doses of alcohols. In the account of the method the exact account of the circulating fluid is not stated and we would not know how strong a percentage the circulating blood contained of alcohol, if we were not told (p.343) that diluting the poisoned circulation with its own bulk (100 c.c.) of saline 0.8% caused quick recovery. Here we learn that the quantity of the solution of blood first used in each experiment was 100 c.c.

The fact that dilution of the poisoned blood as stated above caused quicker recovery, seems to show that it is not the entire quantity of alcohol which flows through the heart in the whole experiment which ought to be taken into account, so much as the percentage strength which flows through heart at any one time. As the quantity of the circulating fluid was 100 c.c. we may arrive at the average percentage strength of the solutions of Ringer and Sainsbury by determining what per cent the various amounts of the different alcohols which they found necessary to arrest the heart, are of the whole solution (100 c.c.). Thus we find that sol.-

tions of the following strength were used:

Methyl	Ethyl	Propyl	Butyl	Amyl
12.1%	6.73%	3.5%	1.%	.4%

Solutions as strong as these used upon isolated dog's hearts would cause instantaneous death. Calculating the weight of the number of c.c. of various alcohols used by Ringer and Sainsbury in grams (from the specific gravity tables of Landolt & Bornstein) the following number of grams are found to have been employed in each case.

Methyl alcohol	9.6318	Grams.
Ethyl "	5.3413	"
Propyl "	2.8393	"
Butyl "	.8242	"
Amyl "	.4148	"

To show the exceedingly large doses used by Ringer and Sainsbury we suffix for sake of comparison the table of Dujardin Beaumetz and Audige:

Alcohols	subcutaneous (pure) fatal dose grams.	death in hrs.	subcut. with Glycerine		By stomach with Glycerine	
			fatal dose in grams.	death in hours	fatal dose in grams.	death in hours
Methyl C H O	6-9.0	30-48	7.2	24	5.5-6.5	12-15
Propyl C H O	4-4.5		3.0-3.6	24-36	3.0-3.4	12
Butyl C H O	2-2.3	6-7	1.92	25	1.76	
Amyl C H O	1.8-2.3	2-7	1.3-1.63		1.4-1.55	3-10

Methyl alcohol according to Dujardin Beaumetz and Audige is stronger than Ethyl 5 grams of pure CH_4O and 7 grams of diluted CH_4O being sufficient to produce death in 24 hours in a dog weighing one kilogram. By comparing the tables of Ringer and Sainsbury with those of Dujardin Beaumetz it is evident that the former used solutions of alcohols on frogs which were far stronger than those required to kill dogs of weight of one kilogram in 24 hours.

It might be urged just here that possibly the alcohols have a far less powerful effect on cold blooded animals than on warm blooded ones, which perhaps might explain the discrepancy. To determine this we injected solutions of various alcohols of the strength used by Ringer and Sainsbury into the jugular veins of frogs and found that without exception the heart had ceased beating in periods of time varying from 5 - 15 minutes.

This observation had been made long before on frogs with Amyl alcohol by Bergeron & L'Hote (Compt rend CXI No. 7) also Gaz heb'd de med N. 35 p.568. These investigators in trying to show the detrimental effects of amyl alcohol used in the Erdmann Uslar method of separation of Morphine. A narcotic extract made by help of amyl alcohol prove that considerable quantities of water shaken with amyl alcohol, in doses of a few drops act as a rapid and fatal poison when injected subcutaneously in frogs and similar-

ly on guinea pigs and rabbits when used in doses of a few cubic centimetres.

It is probably true that alcohol acts with less vigor on cold blooded animals than on the warm blooded, perhaps that conclusion might be deduced from the fact which we will attempt to demonstrate later on, namely, that the physiological activity of alcohols increases with the temperature. Nevertheless it seems clear from our own observations on frogs and those of Bergeron & L'Hote that the quantity of the various alcohols employed by Ringer and Sainsbury were probably too strong to give accurate results.

In volume II p. 483 of the "Studies from the Biological Laboratory of the Johns Hopkins University", is contained a very interesting report by our late Dr. J. R. Duggan on the "Influence of alcohols on the conversion of starch by Diastase".

The main idea suggested here is that the relation between chemical constitution and physiological action of alcohol might be studied by the retardation produced by them in the conversion of starch by diastase; i.e. that the amount of any alcohol required to bring about a certain degree of retardation could be taken as a measure of its physiological strength. Duggan also made some experiments on the amount required of each of the first three primary alcohols of the ethylic series, to prevent for a given time the development of bacteria in a standard solution of beef peptones. The activity he found increased with the addition of each CH_2

group, with the exception of Ethylic alcohol, which was found to cause a break in the series, being less strong than methylic. Dr. Durgan suggested that there might be in all alcohols a chemical property to which their action was due, which might be termed "Alcoholicity", just as with acids we usually have "acidity" with bases "alkalinity." As this "alcoholicity" could be counteracted by addition of small amounts of acid, he supposed that this hypothetical property was allied to alkalinity and caused us to undertake three experiments on the isolated dog heart in which the alcohol was neutralized by minute additions (1-2 drops) according to his suggestions) of acetic and hydrochloric acid; with a view of ascertaining whether the effect of the alcohol would not be diminished thereby. The results however were negative; for blood solutions in which both alcohol and acid were present seemed to be more harmful to the heart than that which contained merely alcohol.

Dr. F. S. Lee in the III volume of this Journal has shown that alkalies in the circulation produce vascular constriction in contradistinction to neutral salts and particular acids which produced dilatation.

It seems certain from observations of many that the alcohols act as vascular dilators, their physiological action therefore is not analogous to that of alkalies, nor is it plausible that it is due to the same property (viz: alkalinity).

This leads us to say a few words concerning the Chemis-

try of Alcohols.

According to differences in their molecular structure as manifested by differences in their conduct upon oxidation, alcohols are divided into primary, secondary and tertiary. In studying the relation between physiological activity and chemical constitution it would be necessary that all alcohols used should belong to the same class.

The alcohols used in the series of experiments to be presently described were all fermentation alcohols, they were all primary.

These alcohols are normal, Methyl, Ethyl and Propyl alcohol, Isobutyl alcohol, and Amyl alcohol of fermentation, which as was shown by Pasteur consists of two isomeric alcohols which both are primary.

The chemical differences between two primary representatives of the same alcohol, are much less than between a secondary and a primary, or between a tertiary and a primary, or between a secondary and a tertiary; so that, as the normal butyl and amyl alcohols cannot be obtained except with difficulty, they are chemically most nearly represented by their primary isomerics. This fact can perhaps be illustrated more clearly by representing the graphic or structural formulae of the first five members of the Ethylic alcohol series.

I.	Methylic	H (COH)
		(H_2)
II.	Ethylic	H C (COH)
		H_2 (H_2)
III.	Propylic	H C C (COH)
		$H_2 H_2$ (H_2)
IV.	Butylic	H C C C (COH)
		$H_2 H_2 H_2$ (H_2)
V.	Amylic	H C C C C (COH)
		$H_2 H_2 H_2 H_2$ (H_2)

It is evident that each formula in this series differs from its preceding one by the group $C H_2$, it is also evident that this addition does not imply derangement each $C H_2$ thus interpolated enters into combinations precisely similar, and hence it must be apparent that whatever change in properties the addition of the $C H_2$ group, effects in the first instance, a like change must be

effected by the addition of the group CH_2 in the second instance, and so on. One is dealing here simply with a chain of like material and structure but composed of more or fewer links. There is, however, one other point for mention, viz: that such a series, whilst it means quantitative difference implies qualitative likeness.

The secondary and tertiary alcohols cannot enter into such a series for comparison, for the additional CH_2 interpolated does not enter into precisely similar relations, but comes into relation with the group $\genfrac{}{}{0pt}{}{COH}{H_2}$ placed in brackets above and so modifies the general structure.

Finally the isoprimary alcohols do not strictly fit into the above series; but the additional CH_2 groups do not modify the arrangement of the group $\genfrac{}{}{0pt}{}{COH}{H_2}$ which is the more important group of the alcohol molecule hence the primary isomerics showing much smaller differences may without much stretch be used for comparison. (Ringer & Sainsbury)

Influence of Alcohol on the Constituents of the Blood.

In considering the physiological action of alcohol, the question of its influence on the plasma and corpuscles of the blood arises.

Not all of the alcohol mixed with the blood in circula-

tion experiments, enters the system and exerts its detrimental influence directly as alcohol; it is probable that a portion of it is used up or changed on coming in contact with the blood constituents. This seems to be confirmed by the observations of "Buchheim" (Archiv der Heilkunde 1866, p. 124) which show that if alcohol is added to fresh blood outside of the body, it is so changed that even a short time afterwards very little of it can be regained. According to the same author, this is in some way a vital phenomenon of the blood, for blood which has been carefully kept from eighteen to twenty four hours, disposes of much less alcohol.

That alcohol is very quickly changed or gotten rid of somehow in the blood was shown by Prof. Martin in experiments reported in previous number of (this Journal) *the Reports from Bio. Lab of J.H*

Lallemand, Perrin and Duroy reported that alcohol accumulated in the blood, and more particularly in the brain and liver, according to J. W. Warren (Harvard) their figures concerning the amount regained from the liver, after administering large doses of alcohol are not convincing, but in the blood only 11% was found and in the brain not more than 1%.

Anstie (Final Experiments etc) gave a small dog one ounce (198.9 grains) of brandy daily for ten days; the amount excreted on the tenth day was only 1.13 grains of alcohol. On the eleventh day the animal was given half an ounce of brandy, killed two hours later, and immediately cut up into small pieces and

every portion of it put into water for extraction.

It appears that only twenty four grains of alcohol were recovered, or not more than one fourth of the amount which had been taken a few hours before.

It seems plausible then that when alcohol reaches the blood directly or through the alimentary canal, a portion of it at least and probably the largest is changed to something else, whatever this is we are not able to tell in the present state of our knowledge.

The direct action of alcohol on the blood constituents has been studied by "Schmiedeberg" (Petersburg Medic Zertschrift 1868 B 14, S 92; also Virchow's Archiv. Vol. 51 p. 171) Here we are informed that blood mixed with alcohol does not give its oxygen up in the presence of a reducing agent as readily as the same blood in a pure state. As it is the role of haemoglobin to yield oxygen which has been taken up by the lungs to oxidizable blood and tissue constituents in the systemic circulation, the action of alcohol on blood just mentioned must be of great importance for processes in the living organism. A direct chemical influence of the alcohol, in addition to its action on heart and nerves, is therefore very probable that is providing alcohol acts on the blood inside of the body in the same way it does outside.

Herrman (Archiv. f. Anat u Physiolog   Reichart u Du-

bois-Reymond 1866 S 27.) describes the effect of the vapor of alcohol on blood thus: The rouleaux of corpuscles break asunder, the latter become spheroidal; these minute spheres become paler and paler and finally disappear entirely, while the plasma colors intensely red and separates out crystals.

According to Maroaud (L'alcool son action physiologique etc.- Paris 1872, p.14) alcohol produces a disturbance between the endos and exosmatic relation between the corpuscles and the serum whereby the nutrition of the former is interfered with.

The following interesting observations were made by Joillet and Hayem of which we find an abstract in Virchow and Hirsch-Jahresbücher d. Gesammt. Medizin 1884, p. 369.

In animals into which alcohol had rapidly penetrated from the stomach, these observers confirm an extensive alteration of the red blood corpuscles; scarcely one third of them was intact, the remaining were partly of the mulberry type, with yellow precipitates of haemoglobin in their interior. They were also partly reduced in size and deprived of their haemoglobin entirely. On injecting alcohol directly into the blood, changed and destroyed corpuscles were found, but no mulberry forms. On introduction of gradual small doses the haemato blasts and leucocytes were strikingly increased, so too the coagulability of the blood. Blood-gas analysis in animals, under not quite intoxicating doses of al-

cohol showed reduction in the respiratory capacity of blood-corpuscles with considerable increase of CO_2 in the blood, which latter fact Joillet partly brings into connection with the oxydation of the alcohol.

C. H. Schulz (Wirkung des Branntweins in der Trunksucht Hufeland's Jourland fur pract Heilkunde 1841 April) has already described the separation of the haemoglobin from the corpuscles and the coagulation of the blood on addition of alcohol in 1841.

Lallemand, Perrin and Duroy (Du role de 'alcool etc. etc. Paris 1860) have observed in the blood of alcoholized animals large quantities of oil globules floating on top after blood was drawn; these were however not observable in the blood of the same animals after they had gotten over the effects of the alcohol. According to Baer (Alcoholisms) Magnus Huss and others have observed the same in the blood from the heart and large veins that died in intoxication. An interesting study of the effects of alcohol on febrile animals was made by Manassein. The corpuscles of the blood of such animals are very much diminished in size by the elevated temperature; this is not the case if the animals are given alcohol, when the corpuscles become larger the longer the intoxication lasts and the more the temperature is lowered by the alcohol. All of these observations and experiments on the influence of alcohol on the constituents of the blood part of which we had ample op-

portunity to work over ourselves, when we were enabled to verify
the main results of Joillet and Hayem, by microscopical examination
of alcoholized blood; these observations justify us in concluding
that in circulation experiments on the isolated heart with alcoholized blood, all of the disturbances observed in the function of
the organ, are not due primarily and exclusively to the direct action of alcohol on the heart tissue, but that the action of the
alcohol on the blood itself whereby its nutritive and oxygenating
function is destroyed should enter into consideration.

Synopsis of the Method.

The method employed in this series of experiments
for isolating the mammalian heart, was that which has been termed
by its originator Prof. H. N. Martin the "Baltimore Method". It
is fully described in a previous number of this journal (Vol. II,
p. 213, plate XV). In the proceedings of the Medico-Chirurgical
Faculty of Maryland (April 29, 1883), the author of the method
states that it has been improved upon by his pupils, so that he
could hardly tell what part of the method was his and what the de-
vice of some one or other of them. And so in this case a number
of changes were made which however are of secondary importance and
will not necessitate giving another description of the method.

In a general way it may be said that the method employed

is given in a paper contained in a previous number of this Journal by Prof. Martin and Lewis F. Stevens B. A. (The Action of Ethyl alcohol upon the dog's heart.) A modification was made in the arteries that were connected with the outflow tube.

In these experiments both carotids were connected with a "horse shoe" canula, from which the outflow tube extended to and through the top of the chamber; here it ended in an outflow orifice, which emptied its blood into a funnel, from here the blood could be directed to any one of four large jars made of glass and being converted into Mariotte's bottles. The same (outflow tube) was by inserting a T glass tube connected with a mercury manometer which recorded the pulse on the paper of a Kymograph. The height of the outflow orifice was so selected as to give a mean pressure of 130 - 150 m.m. of mercury measured by the manometer.

The circulating fluid was supplied from four large glass jars, that were converted into Mariotte's flasks. Each flask could by means of stop-cocks be connected at will with a canula that was inserted into vena cava superior. These connections as may be learned from the references, were made on the outside of, and without opening the chamber.

In the canula which was tied in the stump of the superior cava and only 2 inches from the right auricle was inserted a thermometer which indicated the temperature of the blood as it entered the auricle.

The supply bottles stood in larger jars filled with warm water and acting as jackets. The vapor of the hot water from a trough in which the entire chamber rested, continually surrounded everything inside, keeping the blood constantly at any desired temperature.

If it was desired to lower the temperature, the Bunsen burners beneath the trough were extinguished and a large opening in the top of the case uncovered. If a high temperature was desired the opening was closed and more burners lit beneath the trough. Thermometers hung up in various parts of the chamber rarely differed more than one degree C. from that in the vena cava canula.

The animals used were dogs; an attempt was made to get them approximately of the same weight, in which we were not as successful as we desired.

The circulating nutrient liquid consisted of defibrinated fresh calf's blood and in some cases of defibrinated dogs blood. No differences in the effects of the same alcohol was observed on using the two kinds of blood. In two of the Mariotte's bottles were contained two litres of fresh defibrinated dog's or calf's blood to which 500 cubic centimetres of a 0.75 solution of sodium chloride in distilled water had been added. In the remaining two bottles the same amount of blood was contained, but before it was mixed with the sodium chloride solution the alcohol was

added to the latter in such quantity as to make 1-5% of the whole mixture. One fifth of one per cent. was the amount of alcohol contained in all solutions, at all experiments with exception of those with Amyl alcohol, where 1-5% was found after repeated trials to be too strong a solution to experiment with successfully and solutions of 1-10% and 1-20% were employed, in addition to those containing 1-5%.

The same precautions were taken as in the experiments of Prof. Martin and L. T. Stevens; the outflow of any one of the four Mariotte's bottles in one minute not varying more than 3% from that obtained from any of the other three.

Generally three persons took part in the experiment. One attended to the working of the kymograph, a second had control of the turning on or turning off of pure or alcoholic blood and looking at the thermometers and recording the temperature, while a third was engaged in taking measurements of 30 seconds each at the outflow tube, the blood collected in graduated vessels for measurement, was at once, as soon as noted, poured back into the funnel, from where it flowed into the receiving bottle. Our way of determining the strength of any particular alcohol on the isolated dog's heart, was not that of Ringer and Sainsbury, i.e. trying to establish the amount of alcohol required to kill the heart; as the exact moment of death of this organ is a point that can hardly be

decided upon with accuracy.

The method adopted here was to measure the amount of reduction that a particular alcohol effected in the number of c.c. pumped around by the heart in 30 seconds.

When the amount pumped around by the heart in 30 seconds was quite constant in four or five consecutive measurements, the alcohol containing blood was turned on and measurements immediately undertaken, and the reduction produced in the amount pumped around in 30 seconds, as indicated in measurements; taken as a measure of the strength of the alcohol. After the action of the alcohol had been clearly manifested, pure blood containing no alcohol was turned on, when recovery from the effects of the drug as a rule occurred. Those experiments in which no good recovery occurred under pure blood were rejected, as the reduction in the work might have been due to causes independent of the action of the alcohol.

RESULTS.

Explanation of the Tables.

In the tables given, not all the successful experiments made are recorded, since in some the animals used were so much below the others in average weight, that the results obtained from them could not be used in a comparative study. The variations in our results are undoubtedly due in part to variations in age and weight of the dogs used. To show the effects, the increase or de-

crease of temperature was upon the action of the alcohols, there are recorded experiments in which the temperature was gradually raised from the onset, and such in which it was gradually lowered. Record tables of five successful experiments are given under each alcohol. The tables are easily explained. There are five columns used and headed from left to right thus:

(1) Time, (2) Amount pumped, (3) in 30 seconds, (4) Temperature, (5) Kind of blood.

Taking the record of the experiment with Methyl alcohol on April 20th, 1886, we make the following observations:

The heart was isolated completely at 2:15 P. M. In the measurements at 2:16 minutes, 2:20, 2:22 and 2 o'clock 24 minutes, the heart pumped around 198 c.c.- 197 - 197-- and 196 c.c. in 30 seconds respectively. This was done on pure blood, the temperature being 36°C. at the first three measurements and sinking to 35° C. at the fourth. At 2 o'clock 26 minutes the alcoholized blood was turned on, and at 2.27 minutes a measurement was taken which showed that the heart had only pumped around 186 cub. cent. in 30 seconds at a temperature of 35°C. under 1/5% alcoholized blood. As the previous measurements under pure blood gave 196. c.c. in the same time and at the same temperature we may say that at 35°C. a solution of 1/5% Methyl alcohol will cause a fall in the amount pumped around by the isolated dog heart in 30 seconds, from 196 c.c

to 155 c.c. which is a reduction of 10 c.c.

As the alcoholized blood continues to flow through the heart its action becomes manifest in subsequent measurements. Thus at 2 o'clock 29 minutes the amount pumped around sank to 175 c.c. and at 2:32 minutes to 162 c.c.

At 2 o'clock 33 minutes. Pure blood was again turned on and the alcoholic blood turned off; at once we noticed the recovery taking place as shown in the increased amount of the measurements which at 2 o'clock and 40 minutes have reached 194 c.c. in 30 seconds. Here alcoholized blood is again turned on and pure blood turned off. At 2 o'clock and 43 minutes the next measurement thereafter shows a fall in the amount pumped around in 30 seconds to 174 c.c. By this time the temperature has risen to 37° C. As the previous measurement under pure blood at 2 o'clock and 40 minutes was 194 c.c., we may say, that 1/5% Methyl alcohol in defibrinated calf's blood causes a fall, in the amount pumped around by the isolated heart of the dog in 30 seconds, from 194 c.c. to 174 c.c. or a reduction of 20 c.c. at a temperature of 37°C.

Similarly we observe at 3 o'clock 3 minutes a reduction from 170.5 to 138 c.c. or diminution in amount pumped around at 39° C. of 38 c.c.

gether we obtain the following table: 1/5% solution of Methyl alcohol in defibrinated calf's blood causes the following reductions (diminution in work) of the amount pumped around by an isolated dog heart in 30 seconds.

At 35°C. a fall from 190 c.c. to 180 c.c. or reduction of 10 c.c.
" 37°C.   " " 194 " " 174 " " "   20 "
" 39°C.   " " 170 " " 138 " " "   32 "
" 42°C.   " " 142 " " 92 " " "   50 "

Now by adding the reductions of the measurements at 35° 37° and 39° we get 62, dividing this by 3 gives us the average reduction which is 20.7 c.c. and as the mean temperature of the first three of the above measurements was 37°C.

We may sum up by saying that, The average reduction effected by 1/5% Methyl alcohol, in the amount pumped around by the isolated dog heart in 30 seconds is 20.7 c.c. at 37°C. as evinced from this experiment.

In a similar manner the average reduction in the work done by the heart under the various alcohols was determined, namely: (a) noting the number of cub. cent. pumped around just before the alcoholized blood was turned on; (b) noting the diminution in the number of c.c. immediately after the blood containing the alcohol is turned on; (c) noting the temperature with each measurement. This gave the reduction effected by the alcohol at the various tem-

peratures.

With Methyl Alcohol four experiments were made in which the temperature was gradually increased from the beginning and one experiment in which the temperature was gradually reduced from the onset. The same was the case with all of the remaining alcohols. In determining the general average reduction expressed in cub. cent. which any particular alcohol would effect in the work done by the heart, only those four experiments were made use of in which the temperature was gradually raised from the onset, and that one in which the temperature was gradually reduced from the beginning of the experiment was excluded. Since it was thought more correct to determine the general average reduction (the reduction drawn from all the experiments with the same alcohol) only from experiments in which all conditions were alike.

The temperature (average) at which the general average reduction was in all cases calculated was $37°$ C for reasons to be given later. The weights of the dogs experimented with and the corresponded average reduction which the alcohol produced in the work done by the heart, are given in a separate table

In four experiments with Methyl Alcohol the reductions were

By adding these four figures and dividing the sum by 4 one naturally obtains the mean or general average of the reductions observed in all the experiments under Methyl Alcohol which is 19.48 C.C.

The average weight of the animals used in these four experiments, as may be seen in the table was 15793 grams.

The temperature at which the average is calculated is 37° C.

So one may now derive the follow conclusion.

1/5% Methyl Alcohol in defibrinated blood will at 37° C. reduce the amount pumped around by the isolated heart of a dog weighing 15793 grams--in 30 seconds--by 19.48 cub. cent.

In the same manner the physiological effect of the remaining alcohols was determined. The following figures express the general average reductions of the first four alcohols.

Methyl	$C H_4 O$	19.48	C.C.
Ethyl	$C_2 H_6 O$	17.45	"
Prophyl	$C_3 H_8 O$	79.705	"
Butyl	$C_4 H_{10} O$	131.121	"

When we get to Amyl Alcohol we meet with difficulties. This member of the series proved too powerful to be used successfully in the same strength as the preceeding members namely 1/5%. It was found after repeated trials that solutions of Amyl Alcohol of this strength (1/5%) might be used, if the time in which the

measurements were taken was shortened to 15 seconds, (this period having been 30 seconds in all other experiments), thus considerably abbreviating the time in which the heart was exposed to the poisonous influence of the drug. In two experiments with 1/5% of Amyloxy hydrate in defibrinated calf's blood in which measurements were taken of only 15 seconds duration (May 28th and June 1st, 1896) The general average reduction was found to be 161.16 C.C. at 37° (in 15 sec.)

In four experiments with Amyl Alcohol, however in which solutions of 1/10% were employed and measurements taken in 30 seconds the general average reduction was 161.503 C.C. which is strikingly near the figure given above. So much so, that one might believe, the same effect could be obtained, if one would double the strength of his solutions, but half the time of exposure, or half the strength of the solutions and double the time of exposure. For in the experiments given 1/5% Amyl Alcohol caused the same reduction in 15 seconds, that 1/10% produced in 30 seconds. Whether such a relation as represented really exists we are not able as yet to say. More experiments on this point especially, are necessary.

One experiment was undertaken with only 1/20% solution of Amyl Alcohol. Here the reduction observed in measurements of 30 seconds was 81.33 C.C. This is just one half (or very nearly so) the reduction effected by 1/10% solutions of this alcohol in 30 seconds, or one half the reduction which 1/5% Amyl Alcohol

caused the 15 second measurements. So far we have the following average reductions observed under Amyl Alcohol, with solutions of varying strengths.

Amyl Alcohol Solutions of	Reduction in measurements of 15 seconds.
1/5%	161.16 C.C.
	measurements of 30 seconds.
1/10%	161.508 C.C.
	measurements of 30 seconds.
1/20%	81.33

What the effect of a 1/5% solution of Amyl Alcohol would be shown to be in measurements of 30 seconds, we cannot definitely give, except by speculation. It would be as one might surmise from above table about twice the reduction of 1/5% in 15 seconds, or about twice the reduction of 1/10% in 30 second measurements, or about 323. C.C. reduction in the amount pumped around by the heart in 30 seconds.

This conjecture explains in a theoretical way, why we have not succeeded in keeping 1/5% solutions of Amyl Alcohol flowing through the heart for 30 seconds. Because as we conjecture that would produce a reduction in the amount pumped around by the heart of 323. C.C.

None of the hearts that were isolated by us, however, pumped around such an amount of blood in 30 seconds, and hence the effect of 1/5% Amyl Alcohol in 30 seconds measurements could not be shown.

The Physiological effects of the various alcohols according to our experiments may be therefore expressed by the reductions they produce in the work of the isolated mammalian heart, thus

Methyl	CH_4O	19.46	C.C.
Ethyl	C_2H_6O	17.45	"
Propyl	C_3H_8O	79.705	"
Butyl	$C_4H_{10}O$	161.121	"
Amyl	$C_5H_{12}O$	323.	" (as explained above)

In the aforegoing table it is convenient to state the power of Amyl Alcohol to reduce the work of the heart by 323, as we have determined before, rather than the reduction of 161.508 C.C. which is the amount of reduction it was shown to produce in 10% solutions as determined by 30 second measurements. For by this means we may make clearer the following conclusions.

I. That the physiological activity increases with the complexity of the molecule with exception of the second member viz: Ethyl alcohol which is weaker than Methyl Alcohol.

We have been in possession of these figures for a considerable time, before another definite relation was found to exist among them. It was not so evident before we were able in some way to establish the physiological strength of Amyl Alcohol. The relation between the figures we refer to is this.

That with exception of Ethyl alcohol the figures approach

a geometric progression. Excluding ethyl alcohol which is weaker than Methyl we have the following figures:

CH_4O C_2H_6O C_3H_8O $C_4H_{10}O$ $C_5H_{12}O$
19.46   79.705   161.121   323.

Judging from this, the strength of Propyl alcohol C_3H_8O is very nearly four times that of methyl. Butyl alcohol $C_4H_{10}O$ is twice as powerful as Propyl and Amyl alcohol twice as strong as Butyl alcohol. Putting the results in another way we might say that Amyl alcohol is 16 times as poisonous as Methyl.

Butyl   "   "   8   "   "   "   "   "
Propyl "   "   4   "   "   "   "   "
Ethyl   "   "   less poisonous than   "

Let us assume that the figure for Methyl alcohol had been 20 instead of 19.46 then the remaining alcohols would rank as follows:

Methyl	Ethyl	Propyl	Butyl	Amyl
CH_4O	C_2H_6O	C_3H_8O	$C_4H_{10}O$	$C_5H_{12}O$
20	17.45	80.	160.	320.

II. Another principal conclusion which is easily arrived at on examining the Summary given at the end of each table, on each experiment, is that the physiological activity increases as the temperature increases.

In the experiment with 1-5% Methyl alcohol on April 20th we may observe this fact in the summary.

       temp.
At 35°C a reduction of 10 c.c.

temp.
At 37°C a reduction of 20 c.c.
" 39° " " " 32 c.c.
" 42 " " " 50 c.c.

The same fact can be demonstrated in each table. In such experiments in which the temperature was gradually lowered from the onset (as in the following with Methyl alc. on April 23rd with Ethyl alc. on May 2nd; with Amyl alcohol on June 4th, '86) it can be demonstrated that the physiological activity decreases as the temperature decreases. Thus in the experiment with Methyl alcohol on April 23rd '86 we have the following results.

At 40° reduction of 31. c.c.
" 39° " " 26. "
" 38 " " 24.5 "
" 37° " " 20.5 "
" 36° " " 17.5 "

III. A third main conclusion that can be derived from our results, is that the hearts of heavy animals are less affected by the various alcohols than those of lighter animals. Thus, in the following tables the weights of the different animals used are given, together with the amount of reduction any particular alcohol was capable of producing in the number of cub. cent. pumped around by the isolated hearts of those animals.

Weight of the Animals used, and corresponding reductions in the amount pumped around by the isolated hearts of these animals in 30 seconds at a temperature of 37° C.

I. Methyl alcohol

	Weight of Dogs	Reductions
Experiment April 20 '86 = 15000 Grams		20.7 c.c.
" " 21 " = 15983 "		19.583 "
" " 27 " 16770 "		18.583 "
" " " " = 16420 "		19. "
Average weight	15793 "	Average reduction 19.46 c.c.

II. Ethyl alcohol

	Weight of dogs	Reductions
Experiment May 4, '86 = 14605 Grams		20.16 c.c.
" " 6, " = 15480 "		15.83 "
" " 7, " = 15005 "		15.5 "
" " 10, " = 14090 "		18.33 "
Average weight	14795 "	Average Reduction 17.45 c.c.

III. Propyl alcohol

	Weight of Dogs	Reductions
May 18th, '86 No. I 16510 Grams		75. c.c.
" " " " II. 15095 "		80.83 "
" 20th, " " I 15165 "		78.83 "

May 20th, '86 No. II   17430 Grams   78.83 c.c.
                        ------   ------
            Average weight 16050 "   Average Reduction 79.705 c.c.

            IV.   Isobutyl Alcohol
                  Weights of dogs   Reductions
May 24th, 1886 No. I   17406 Grams   161.66 c.c.
 "   "   "   " II   17450   "   160.66   "
 "   25th "   15925   "   162.166 "
 "   26th "   17500   "   160.   "
                        ------   --------
        Average Weight 14570   "   Average Reduction 161.121 c.c.

            V.   Amyl Alcohol
                  One twentieth per cent. 1-20%
June 2nd, '86 Weight   Reduction in 30 sec.
              19416 Grams   81.33 c.c.
                  One tenth per cent. 1-10%
       Weight of dogs
Experiment June 3rd, '86 No. I   16450 Grams   163.366 "
     "   "   "   "   " II 17000   "   163.83   "
     "   4th,   "   17235   "   159.83   "
     "   May 28th "   18015   "   159.   "
                                  --------   --------
Average weight   17175   "   Average Reduct. 161.508 c.c

Amyl Alcohol 1-5%

	Weight of dogs	Reduction in 15 sec.
Experiment May 28th, '86	17600 Grams	161.66 c.c.
" June 1st, "	18112 "	160.66 "
Average weight	17856 "	Average Reduct. 161.16 c.c.

For reasons given elsewhere 15% Amyl alcohol could not be turned on for 30 seconds without killing the heart. In two experiments in which the alcohol was used in solutions of this strength, the observations were taken in periods of 15 seconds. It is seen that there exists a definite relation between the weight of the animals used and the reduction effected by a certain percentage of the alcohol. Although the attempt was made to obtain dogs of approximately the same weight, there are slight deviations. Upon examining the tables presented, closer, it is apparent that under any given alcohol, the reduction effected in the amount pumped around is greatest with the dogs of least weight and vice versa. From this we may conclude that heavier dogs are less affected by the alcohol than lighter ones.

In the experiments on Amyl alcohol a definite relation becomes evident, between the effects of solutions of different strengths and also between the strength of an alcoholic solution and the time in which it is allowed to act.

For instance, as it is stated in the table, 1-5%; solu-

tions of Amyl alcohol will produce, the same effect in reducing the heart's work, in 15 seconds, which 1-10% of the same alcohol does in 30 seconds. Furthermore a solution of only 1-20% of Amyl alcohol will only effect a reduction of approximately one half that of a 1-10% solution. The number of experiments that were undertaken with Amyl alcohol in which variations occurred in the strengths of the solutions, and in the periods of time in which the latter were allowed to flow through the heart,- is not sufficient to permit us to judge definitely of the effect of these changes. We can therefore not as yet conclude whether, reducing the time of exposure one half, will half the effect produced, or whether doubling the strength of a solution will exactly double the effect. It is not improbable that some such relation exists, which however would only become evident, if one could perfectly control all the conditions that influence these experiments, such as weight and age of dogs, quality of the blood used, temperature and time of exposure etc., etc.

This is perhaps the best place to speak of the difficulties connected with the attempt to determine the effect or power of Amyl alcohol in reducing the work of the isolated heart. With all alcohols except Amyl the strength of the solutions used was 1-5% and the period of time in which measurements were taken was 30 seconds.

tion of 1-5% could not be turned on and allowed to flow through the heart for 30 seconds without causing its speedy death. To arrive at a conclusion concerning the effect of Amyl alcohol, that might be available in a comparative series with the preceding alcohols, two things suggested themselves, (1) either the time in which the solution of 1-5% Amyl alcohol was turned on through the heart, had to be reduced, or (2) the strength of the solution itself. Both of these methods were tried with results afore stated.

It appears from these results that in four experiments in which solutions of 1-10% Amyl alcohol, were allowed to flow through the isolated heart, the average reduction effected in the number of cubic centimetres pumped around in 30 seconds was 161.508 c.c. at a temperature of $37°$ C. The average weight of the dogs being 17175 grams.

In two experiments, that were conducted with solutions containing 1-5% of Amyl alcohol, the average reduction effected in the work done by the heart, determined in measurements of 15 seconds duration, was 161.16 c.c. at $37°$C. The average weight of the dogs being 17856 Grams.

In one experiment in which a solution of 1-20% Amyl alcohol was used, the (average) reduction in the amount pumped around by the heart in 30 seconds amounted to 81.33 c.c. at $37°$, the weight of the dog used was 19416 Grams. Although we have no re-

sults showing the effect of a solution of 1-5% Amyl alcohol in measurements of 30 seconds, the above figures suffice to demonstrate that, as regards its toxic effect Amyloxyhydrate fits into the series with the other alcohols, being twice as strong as Butylic alcohol, which is twice as strong as Propylic. The position of Methyl and Ethyl alcohol in this progressive series has been considered elsewhere.

As the physiological activity of the various alcohols increases with the complexity of the molecule, so other properties increase, namely, the vapor density, the specific gravity and the boiling point becomes higher, this becomes evident from the following table:

Name of Alcohol	Chemical Structure	Vapor Density $H_2=1$	Specific Gravity $H_2O=1000$	Boiling Point.
Methyl	CH_4O	16	.8142	55.1° C
Ethyl	C_2H_6O	23	.80625	78.3°
Propyl	C_3H_8O	30	.8198	98.
Butyl	$C_4H_{10}O$	37	.8242	117.°
Amyl	$C_5H_{12}O$	44	.8296	135.

The vapor density of these alcohols increases 7 for every increment in composition of CH_2

The boiling point increases on the average 19.6 for every increment of CH_2. There is also a constant increase of the specific gravity as the complexity of the molecule increases, this however occurs in the second or third decimal point, and Ethylic alcohol forms an exception to the statement.

POSITION OF ETHYLIC ALCOHOL IN THE SERIES.

It is evident from these experiments and those of Dujardin, Beaumetz and Audige and J. R. Duggan, that ethyl alcohol causes a break in the series being less hurtful than methyl. Since ethylic alcohol is probably constantly present in minute quantities in the atmosphere, as fermentation is going on in many ways and places, and as dogs probably always get some little ethyl alcohol in their food, such as in the bread and in fermented meat, this fact has been explained theoretically by supposing that an organism can establish a certain degree of tolerance of one alcohol and retain its sensitiveness to others.

A further explanation of the tolerance of ethylic alcohol by certain animals, is furnished by a fact which is becoming more and more established, and that is: that ethyl alcohol in minute quantities is a normal constituent of almost all animal tissues. One of the clearest investigations on this subject is published by A. Rajewski in Phlügers Archiv. (B XI . 122, 1875) who

obtained the iodoform reaction in the distillate of fresh Liver, Brain and muscles of animals perfectly free from alcohol. Rajewski concluded that these tissues either always contain certain quantities of ethyl alcohol, or that the iodoform test is not a sufficient proof of the existance of ethyl alcohol. C. Binz (Vorlesungen uber Pharmakologie Vol. II p. 357) holds a similar view, according to which this test indicates alcohol in traces even, but also carbohydrates, albumen, fibrin casein and glue. According to L. Millon (Compt. rend 1845 Vol. 4, p. 828) the iodoform reaction has been obtained with saliva of teetotalers.

This reaction as is perhaps well known is carried out by adding a few drops of a solution of Iodine in Iodide of potassium to the solution to be examined, then just enough Sodium hydroxide is added to make the yellow color disappear, then the mixture is heated up to about $60°$, - on cooling the solution becomes a yellowish tint--and the minutest trace of alcohol will cause the formation of iodoform which can be recognized under the microscope by its hexagonal crystals. Reaction in short is the following
$8I + C_2H_6O + 6 NaOH = 5H_2O - Na CHO_2 + 5 Na I + CHI_3$.

Although this reaction is not reliable in testing for alcohol in living tissues, it has been satisfactorily shown that these do contain alcohol. One of the chief workers in this field is J. Bechamp (Paris)

In the Compt. rendu T. LXXXIX, p. 573 (Sur la presence de l'alcool dans les tissue animaux pendant la vie et apres la mort), Bechamp states that he extracted 0.8 Grams of alcohol from a piece of meat weighing 6 lbs., that he has found alcohol in sheeps liver immediately after death, also in human brain. He points out a connection between vital phenomena and phenomena of decomposition under the influence of Mikrozymes. This investigator has also found alcohol to be a normal constituent of fresh milk (Journ Chem. Soc. vol. 26, p. 763). He has developed a theory concerning fermentations intra vitam which is based on what are termed by him the Microzymes. These are described by him as being the physiologically active parts of cells of animal tissues, they are minute granular masses, that have the power of inducing fermentation like other cellular ferments, but differ in quality and energy of action according to the part of the organism from which they are taken. The Microzymes then are organized ferments, occupying geneologically an intermediate position between the two typical classes of organized ferments viz: Yeast and Bacteria. They are supposed to be the centre of all chemical changes of living and dead tissues. (Bechamp. Archiv. de Physiol. Normal et Pathol). The microzymes belong to the animal organism as internal parts without having to get into them from the air or other medium they are even in perfect health capable of causing all those processes intra vitam which are commonly known as fermentations and

post mortem all those metamorphoses which are comprised under the phenomena of decomposition. All soluble ferments are products or secretion of microzymes and are called Zymases (A. Bechamp Archiv. de Physiol. 1882 P. 28) Diastase Trypsin and Pepsin are termed Hordeozymase, pancreazymase and gastrozymase.

It will be impossible for us to enter upon the method of J. and A. Bechamp, of isolating these microzymes, which is given in Compt rend LXXXI p. 226-229. Our intention merely was to point out a theory according to which the existence of ethyl alcohol in living tissues, might be explained and also the tolerance of certain animals toward this alcohol. One of the most satisfactory evidences of the actual occurrence of alcohol in tissues is given by Binz (Vorlesungen uber Pharmakologie). Parts of the animal organism, on distillation in a tightly closed vessel and treating the distillate with finely divided platinum, gave an acid liquid which reduced a warmed ammoniacal solution of silver---consequently it must have contained aldehyde.

    Cause of the diminution in
    the work of the isolated
    heart under alcohol.

the blood vessels or on the muscular tissue of the heart has received proper attention. It appears that the main effect of the drug is exerted upon the cardiac muscular tissue. It is probable that the blood vessels are also acted upon to a certain extent, perhaps that the alcohol causes a vascular dilatation leading on to engorgement of the heart with the poisoned blood, destroying the tonicity of the blood vessels and the contractility of the ultimate muscle fibrils. The great swelling of the heart has been described in the article above, and its relation to the character of the heart beat, the main feature of which is the imperfect systole so that the ventricular cavity is not obliterated at the end of the systolic contraction, and the less so the longer the alcohol has been administered. This incomplete systole is compensated for by a more extensive diastole as long as the heart can swell. As soon as the pericardium prevents this excessive expansion in diastole, the difference between diastolic and systolic capacity becomes less and less and the heart pumps around less blood. An interesting matter in this connection is a fact which we have observed in several cases (6) in which the heart was expanded and swollen to a great extent under alcohol and that is, that by placing a stethoscope directly on the organ, we could hear plainly so called regurgitant murmurs. These murmurs are undoubtedly on both sides of the heart. Perhaps they are caused by the swelling and expanding of the heart excentrically, whereby the fibrous rings

to which the valves are attached are so enlarged that the valves are no longer large enough to cover them, and there is possibly a condition of things known as tricuspid and mitral insufficiency and regurgitation. In some of the dogs hearts that were filled with water after the experiments, although they had collapsed to some extent the valvular insufficiency was quite apparent. This fact was also discovered by freezing a heart that had just been experimented upon, and ligating it so that none of its contents could escape. On dissecting away the auricles from the frozen organ it could be seen that the valves did not entirely cover the orifice. In all experiments we noticed small hemorrhages into the cardiac tissue, these were very considerable with the higher alcohols and partook of the nature of ecchymoses. It was thought that they were the direct result of the action of the alcohols, by injecting considerable quantities of alcohol into the jugular veins of dogs however, we were in no instance able to produce them on the heart.

The reason why all averages were calculated at 37 degrees C. was that it was found that this temperature could be most easily maintained and principally because the averages could not have been drawn at a higher temperature on account of the increase in poisonous action particularly of Amyl alcohol, which in the strength used by us, invariably produced rapid death of the heart above 38. No change in the pulse rate was noticed under alcohol, in any of

ents.

The writer wishes to express his indebtedness to Prof. Martin for much good advice and many useful suggestions.

BIOGRAPHICAL NOTE.

The author of this Thesis was born in Baltimore in 1863 and received his preliminary education in the public schools of that City. From 1871-1878 he studied at the Royal Gymnasium at Hanau and Wiesbaden, Germany. In the latter City he was also a student of the Conservatory for three years. Returning to America, he graduated at the Bryant, Stratton & Sadler Commercial College in 1880. In the Baltimore City College in 1882, and as Doctor of Medicine at the University of Maryland in 1884.

He was assistant resident physician to Bay View Asylum for one year. Was accepted as Candidate of Philosophy by the Johns Hopkins University in 1886, having previously held a Hopkins scholarship and subsequently represented the management of the University at the Baltimore Insane Hospital for two years as physician in charge. His chief studies have been Physiology, Chemistry and Psychology. His main advisers were Professors H. Newell Martin, Ira Remsen and G. Stanley Hall; finally, he wishes to conclude with an expression of deep gratitude toward these gentlemen.

EXPERIMENT with 1/5% C. H$_4$ O Methyl Alcohol; April 20th, 1886.
Dog under Morphine and Curare. Defibrinated Calf's blood used.
Heart isolated at 2:15 min. Remained in good condition 1 hr.17 min.

Time	Amount pumped around in 30 sec. at Temp.			Kind of blood
2:18	196 C.C.	30 sec.	36°	Pure
2:20	197 "	"	"	"
2:22	197 "	"	"	"
2:24	197 "	"	35	"
2:26				C H$_4$ O 1/5%
2:27	186 "	"	35	"
2:29	175 "	"	35.5	"
2:32	162 "	"	36	"
2:33				Pure
2:34	174 "	"	36	"
2:36	185 "	"	36.5	"
2:38	194 "	"	36.5	"
2:40	194 "	"	37	"
2:42				C H$_4$ O 1/5%
2:43	174 "	"	37	"
2:45	160 "	"	37	"
2:47	148 "	"	37.5	"
2:50	102 "	"	38	"

Amount pumped around in 30 sec.	at	Temp.	Kind of Blood
			Pure
130 C.C.	30 sec.	38.5°	"
153.5 "	"	38.5	"
159.75 "	"	39	"
163.75 "	"	39	"
170 "	"	39	"
			CH_4O 1/5%
138 "	"	39	"
130 "	"	39.5	"
114 "	"	40	"
			Pure
120 "	"	40	"
132 "	"	40	"
141.5 "	"	40.5	"
142 "	"	41	"

Action of CH_4O 1/5% increases with the temperature. Thus: at 35
Cent. We observe a fall from 196 C.C. to 186 C.C. or 10 C.C. in 30 sec
at 37 "   " " "   " 194 " " 174   " 20 C.C. " " "
" 39 "   " " "   " 170 " 138   " 32 C.C. " " "
" 42 "   " " "   " 142 " 92   " 50 C.C. " " "

This gives an average of 20.7 C.C. reduction of amount pumped around under 1/5% CH_4O at Temp. 37° C. in 30 sec. measured one minute after CH_4O 1/5% blood was turned on, and calculated from the average of the measurements at 35° C 37° C and 39° C

EXPERIMENT with Methyl Alcohol; April 21st, 1886.

(a) Defibrinated Calf's blood
(b)   "   "   " with 1/5% CH_4O

Heart isolated 3. P. M.   Heart lived 1 hr. 5 min.

Time	Amount pumped in 30 sec.	at Temp.	Kind of blood.
3:3	108 C.C. "	$34.5°$	Pure
3:4	110 " "	34.5	"
3:5	112 " "	35	"
3:6	112 " "	35	"
3:7	112 " "	35	"
3:9			CH_4O
3:10	98 " "	35	"
3:11	88 " "	35.75	"
3:12			Pure
3:14	102 " "	36	"
3:15	108.5 " "	36	"
3:16	110 " "	36	"
3:17	112 " "	36.5	"
3:18	110.5 " "	37.5	"
3:19			CH_2O
3:21	90.75 " "	37	"
3:23	80.5 " "	37.5	"

Time.	Amount pumped in 30 sec.		at Temp.	Kind of blood.
3:25	72 C.C.	"	37.5²	C H₄ O
3:26				Pure
3:27	91 "	"	38	"
3:28	90 "	"	38	"
3:29	102.5 "	"	38.5	"
3:30	102 "	"	38.5	"
3:31				C H₄ O
3:33	73 "	"	39	"
3:34	68 "	"	39.5	"
3:36	54 "	"	39.75	"
3:37				Pure
3:38	70 "	"	40	"
3:39	75 "	"	40.5	"
3:40	74 "	"	Doors opened 39	"
3:42	72 "	"	Pericardium cut 38.5	"
3:43	80 "	"	38.5	"
3.45	88 "	"	38	"
3:46	94 "	"	38	"
3.47				C H₄ O
3:49	80 "	"	37.5	"
3:50	74 "	"	37	"
3:52	70 "	"	37	"
3:54				Pure

Time.	Amount pumped in 30 sec.	at Temp.	Kind of Blood
3:55	75 C.C. "	37°	Pure
3:56	80 " "	37	"
3:57	80 " "	36.5	"
3:58			C H₄ O
3:59	74 " "	36	"
4: P.M.	68 " "	36	"
4:1			Pure
4:2	70 " "		"
4:3	70 " "		"
4:4	74 " "		"
4:5	68 " "		"
4:7	60 " "		"
4:9	60 " "		"

Heart irregular and beginning to die.

Methyl Alcohol 0.20% in this experiment caused a fall
from 112 to 98 or 14 C.C. at 35° C)
)
 " 110 " 90.75 " 19.75 " " 37°   ) in 30 seconds.
)
 " 103 " 78 " - 25 " " 39°   )
           -----   -----
           58.75 "   111

Average 19.583 C.C. in 30 sec. at 37° C.

1 EXPERIMENT. April 27th, 1886.

Dog under Morphine and Curare. Calf's blood defibrinated and Alcoholized blood 1/5% of CH_4O Methyl Alcohol.
Heart isolated at 2:30 P.M. Heart in good condition for 1 hour and 10 min.

Time	Amount pumped in 30 sec.	at Temp.	Kind of blood.
2:33	110 C.C. 30	35°	Pure
2:35	111 " "	35	"
2:37	112 " "	35	"
2:38			CH_4O
2:39	100 " "	35	"
2:40	98.50	35.75	"
2:42	90 " "	36	"
2:43			Pure
2:45	105 " "	36	"
2:47	108 " "	36.5	"
2:49	110 " "	36.75	"
2:51	108 " "	36.5	"
2:53	110 " "	37	"
2:55			CH_4O
2:56	91 " "	37	"
2:58	85 " "	38	"
2:59	80 " "	38	"

Time.	Amount pumped in 30 sec.	at Temp.	Kind of Blood.
3: P.M.			Pure
3:2	94 " "	38°	"
3:5	95 " "	38	"
3:8	98.5 " "	38.5	"
3:9	(not quite) 99.75 (100 C.C.)	3 .5	"
3:10			CH_4O
3:12	75 C.C.in 30 sec.	39	"
3:14	70 " "	39.5	"
3:15	64 " "	39.75	"
3:16			Pure
3:17	70 " "	39	"
3:18	72 " "	39	"
3:20	68 " "	38.5	"
3:21	(Pericardium cut off)		
3:23	90 " "	38	"
3:24	93 " "	38	"
3:25	95 " "	37	"
3:26			CH_4O
3:28	85 " "	36.5	"
3:29	80 " "	36.5	"
3:30	80 " "	37	"
3:31			Pure
3:33	85 " "	37.5	"

Time	Amount pumped in 30 sec.			At Temp.	Kind of blood
3:35	90	"	"	38°	Pure
3:36					C H$_4$ O
3:37	70	"	"	38.5	"
3:38	60	"	"	39	"
3:39	40	"	"	39	"
3:40	20	"	"	39	"
3:41	stopped			39	"
3:42					Pure
3:43				39	

Heart pumped around again for a few seconds and then stopped very much dilated; heart showed small eccymoses, lungs in good condition.

Notes to Experiment I on April 27th, 1886.

Methyl Alcohol CH_4O in form of 1/5% solution in defibrinated Calf's blood, causes a diminution in the amount of blood pumped around by an isolated dog's heart in a given period of time (30 seconds).

The degree of diminution, i. e. the effect of CH_4O on the heart tissue increases with the temperature, causing a fall from 112 to 100, i. e. a reduction of 12 C.C. at $35°$ C.

" 110 "   90 " " "   " 19 "   " 37 "
"99.75"   75 " " "   " 24.75 " " 39° "

Average fall of 18.583 C.C. at 37.° C in 30 sec. Methyl alcohol then causes an average reduction of about 18 C.C. at a mean temperature of 37° in this experiment.

II Experiment on April 27th, 1886.
Dog under Morphine and Curare. Calf's blood defibrinated.
Calf's blood alcoholized with 1/5% Methyl Alch.
Heart isolated 4:15. Heart lived one hr. and 11 min.

Time.	Amount pumped in 30 sec.	at Temp.	Kind of blood.
4:17	116 C.C. "	40.5 C	Pure
4:18	117 " "	40.5	"
4:19	116 " "	40	"

Amount pumped in 30 sec.	at Temp.	Kind of Blood.
110 C" C. "	40°	Pure
		C H₄ O
80.5 " "	40.5	"
70 " "	40.5	"
		Pure
90 " "	40	"
95 " "	39.5	"
99 " "	39.5	"
105 " "	39.5	"
111 " "	39.5	Pure
115 " "	39	"
116 " "	"	"
115 " "	39	"
		C H₄ O
90 " "	39	"
80 " "	"	"
70 " "	"	"
		Pure
78 " "	38.5	"
90 " "	38.5	"
10.35 " "	38.5	"
108 " "	38	"
107.5 " "	37.5	"

Amount pumped in 30 sec. c.c.	at Temp.	Kind of blood.
		CH_4O
88.5 " "	37°	"
78.5 " "	37	"
68 " "	36.5	"
		Pure
76 " "	36.5	"
80 " "	36	"
80 " "	36	"
78 " "	35.3	"
80 " "	35.5	"
		CH_4O
66.5 " "	35	"
60 " "	35	"
	(Lamp turned on)	Pure
69 " "	35	"
78 " "	35.5	"
80 " "	36	"
80 " "	36.5	"
	(Pericardium cut off)	
83 " "	36.5	Pure
88 " "	36.5	"
90 " "	37	"
		CH_2O

Time	Amount pumped in 30 sec.	at Temp.	Kind of blood.
5:11	76 C.C. " "	37°	C H$_4$ O
5:12	69 " "	37	"
5:13	60 " "	37.5	"
5:14			Pure
5:15	70.5 " "	37.5	"
5:16	75 " "	38	"
5:17	76 " "	38	"
5:18	76 " "	38.5	"
5:19			C H$_4$ O
5:20	60 " "	39	"
5:21	50 " "	39	"
5:22	42 " "	39.5	"
5:23			Pure
5:25	55 " "	39.5	"
5:26	50 " "	39.5	"
5:27	46 " "	40	"
5:28	40 " "	40	"

Annotations to Experiment N 11, Apr.27 1888. Methyl alcohol 1/5%
) Heart began to beat very irregular, pumped a-
) round less and less and finally stopped beat-
) ing in diastole very much dilated with small ec-
) chymoses all over it.

Observations were begun with a comparatively high temperature be-

cause the blood had time to become heated up---this being the sec-

ond experiment of the day. The temperature was lowered by turning down the lamps and opening a large hitherto covered hole in top of case, when the temperature reached $35°$; this was again closed and the lamps turned on.

In this experiment 0.20% CH_4O produced

a fall from 116 C.C. to 86.5 C.C. or 29.5 C.C. at temp. $40.5°$

" " " 115 " " 90   " 25 " " " $39°$

" " " 107.5" " 88.5   " 19 " " " $37°$

" " " 80 " " 66.5   " 13.5 " " " 35

Average fall of 19 C.C. at $37°$ C. in 60 sec. This makes an average reduction in the amount pumped around of about 6 C.C. for every $2°$ of temp. or 3 C.C. for every single degree of increase in temperature.

EXPERIMENT April 23rd, 1886 Methyl Alcohol 1/5% in defibrinated Calf's blood. Dog under Morphine and Curare.

Heart isolated 2:40 P.M. under experimentation for

Time	Amount pumped in 30 second	Temp.	Kind of blood.
2:43	128 C.C. "	40°	Pure
2:44	129 " "	40.5	"
2:45	131.5" "	40.5	"
2:46	132 " "	40	"
2:47			Alcohol 1/5%
2:48	101 " "	40	"
2:49	75 " "	40	"
2:50	48 " "	40	"
2:51	20 " "	39.5	"
2:52			Pure
2:53	45 " "	39.5	"
2:54	76 " "	39.5	"
2:55	105.5" "	39.5	"
2:56	128.5" "	39	"
2:57	132 " "	39	"
2:59			Methyl Alc.1/5%
3:	106 " "	39	"
3:02	82 " "	39	"
3:03	58 " "	38.5	"
3:05	32 " "	38.5	"

Time	Amount pumped in 30 seconds			Temp.	Kind of blood.
3:07					Pure
3:08	60 C.C.		"	38.5°	"
3:09	31.5	"	"	37.5	"
3:10	105	"	"	37.5	"
3:11	123	"	"	38	"
3:12	129	"	"	38	"
3:14	129.5	"	"	38	"
3:15					Methyl Alc.1/5%
3:16	105	"	"	"	"
3:17	31	"	"	"	"
3:18	55	"	"	37.5	"
3:19	30	"	"	"	"
3:20					Pure
3:22	70	"	"	37.5	"
3:25	124	"	"	"	"
3:26	128	"	"	37	"
3:28	128.5	"	"	"	"
3:29					Methyl Alc.1/5%
3:30	103	"	"	37	"
3:32	55	"	"	36.5	"
3:34	24	"	"	36.5	"
3:35				"	Pure
3:37	66.5	"	"	"	"

pumped in 30 seconds	Temp.	Kind of blood.
102 C.C. "	36.5	Pure
124 " "	36	"
127 " "	"	"
127.5 " "	"	"
		Methyl Alc. 1/5 %
110 " "	"	"
95 " "	"	"
78 " "	"	"
60 heart irregular	"	"
		Pure
60 more irregular		"
20 C.C. 30 seconds		"
Heart did not recover under pure blood.		

his experiment the temperature was taken high (40 C)
ng of the experiment and gradually lowered as it pro-
other experiments the effect of the alcohol increased
ture increased---in this case it decreases as the
creases---in both cases showing dependence upon the

The effects of 1/5% Methyl Alcohol as shown by the

At 40° a fall from 132 C.C. to 101 C.C. or reduction of 31 C.C.
" 39° " " " 132 "   " 106 "   "   "   " 26 "
" 38° " " " 129.5" " 105 "   "   "   " 24.5"
" 37° " " " 128.5" " 108 "   "   "   " 20.5"
" 36° " " " 127.5" " 110 "   "   "   " 17.5"

Heart stopped in diastole, very much swollen and covered with minute ecchymoses.

The average taken from above measurements at 36° 37° and 38° is 20.83 C.C. reduction in amount pumped around in 30 seconds at 37°.

Average reduction of all (4) Experiments with 1/5% Methyl Alcohol (excluding the one on April 23rd, 1886)

April 20th, 1886--20.7 C.C. reduction at 37°
"   21st, 1886--19.583 "   "   " "
I. " 27th, 1886--19.583 "   "   " "
II. " 27th, 1886--19.   "   "   " "

4)77.866(19.46 C.C.= Total average reduction of amount pumped around in 30 seconds at 37° C. by isolated dog's heart. (calculated from results of four experiments.)

EXPERIMENT, May 2nd, 1896, Ethyl Alcohol 1/5% in defibrinated calf's blood. Dog under Morphine and Curare. Heart isolated 2:24 P.M. under experimentation 1 hour and 44 minutes.

Time	Amount pumped in 30 seconds			Temp.	Kind of blood.
2:28	164 C.C.	"		42°	Pure
2:30	160	"	"	"	"
2:32	166	"	"	"	"
2:33	168	"	"	"	"
2:34					Ethyl Alc. 1/5%
2:35	120	"	"	"	"
2:36	90	"	"	"	"
2:37	60	"	"	41	"
2:38	24	"	"	"	"
2:40					Pure
2:42	65	"	"	"	"
2:44	104	"	"	"	"
2:46	148	"	"	"	"
2:48	165	"	"	"	"
2:50	168	"	"	"	"
2:52				"	Ethyl Alc. 1/5%
2:53	130	"	"	"	"
2:55	98	"	"	"	"
2:56	69	"	"	40	"

Amount pumped in 30 seconds			Temp.	Kind of blood
40	"	"	40°	Ethyl alc. 1/5%
				Pure
34	"	"	"	"
120	"	"	"	"
124	"	"	"	"
105	"	"	"	"
130	"	"	"	"
				Ethyl Alc. 1/5%
103	"	"	"	"
100.5	"	"	39	"
70	"	"	"	"
57	"	"	"	"
				Pure
75	"	"	"	"
110	"	"	"	"
153	"	"	"	"
105	"	"	"	"
125.5	"	"	"	"
				Ethyl Alc. 1/5%
137.5	"	"	"	"
105	"	"	"	"
79	"	"	38	"
				Pure

Time	Amount pumped in 30 seconds	Temp.	Kind of blood.
3:38	108 C.C. "	38°	Pure
3:40	138 " "	"	"
3:42	104 " "	"	"
3:44	105 " "	"	"
3:45			Ethyl Alc. 1/5%
3:46	142 " "	"	"
3:47	120 " "	"	"
3:48	96 " "	"	"
3:49			Pure
3:50	124 " "	37	"
3:52	146 " "	"	"
3:54	159.5 " "	"	"
3:56	160 " "	"	"
3:58	162 " "	"	"
3:59	105 " "	"	"
4:		"	Ethyl Alc. 1/5%
4:01	140.5 " "	"	"
4:02	120 " "	"	"
4:03	99.5 " "	36	"
4:04			Pure
4:06	124 " "	"	"
4:08	148 " "	"	"
4:10	160 " "	"	"
4:11	162 " "	21	"

Time	Amount pumped in 30 seconds	Temp.	Kind of blood.
4:12			Ethyl Alc. 1/5%
4:13	145.5 C.C. "	36	"
4:14	128 heat irregular	"	"
4:15	110 C.C. 30 seconds	"	"
4:16	40 Kymograph tracing faint.	"	
4;17	heart did not recover under pure blood.	"	Pure

RESULTS:

At 42°	a fall from	168	C.C.	to	126	C.C.	or	red.	of	42	C.C.	) In
" 41	" "	"	163	"	" 130	"	"	"	"	33	"	)amount
" 40	" "	"	166	"	" 133	"	"	"	"	33	"	)pumped
" 39	" "	"	165.5	"	" 137.5	"	"	"	"	28	"	)around
" 38	" "	"	165	"	" 142	"	"	"	"	23	"	)in 30
" 37	" "	"	165	"	" 148.5	"	"	"	"	16.5	"	)sec-
" 36	" "	"	162	"	" 145.5	"	"	"	"	16.5	"	)onds.

As the temperature sinks not only the amount pumped around becomes less, but also the reduction produced by 1/5% Ethyl Alcohol (become less and less). The temperature was intentionally made to be high at onset and then gradually lowered. Average of Experiment on May 2nd, 1886 with 1/5% Ethyl Alcohol in defibrinated calf's blood 1933 C.C. reduction in amount of C.C. pumped around in 30 seconds at 37^c

EXPERIMENT March 4th, 1896, Ethyl Alcohol 1/5% in defibrinated dog's blood. Dog under Morphine and Curare. Heart isolated 1:24 P.M. remained good for 1 hour and 04 minutes.

----------:---------

Time	Amount pumped in 60 seconds	Temp.	Kind of blood.
1:26	130.C. "	35°	Pure
1:28	138 " "	"	"
1:30	131.5" "	"	"
1:32			Ethyl Alc. 1/5%
1:33	130 " "	"	"
1:34	121.5" "	35.5	"
1:35	112 " "	"	"
1:36	100 " "	"	"
1:37			Pure
1:38	110.5" "	"	"
1:39	113 " "	"	"
1:41	130 " "	36	"
1:43	137.5" "	"	"
1:44			Ethyl Alc. 1/5%
1:45	125 " "	"	"
1:47	101 " "	"	"
1:49	90 " "	36.5	"
1:51	84.5" "	"	"

Time	Amount pumped in 30 seconds			Temp.	Kind of blood.
1:52					Pure
1:53	78 C.C.	"		36.5	"
1:54	90	"	"	"	"
1:56	112	"	"	"	"
1:58	126	"	"	"	"
2:	135	"	"	37	"
2:01	136	"	"	"	"
2:03		"	"		Alcohol 1/5%
2:04	115	"	"	37	"
2:05	95.5	"	"	"	"
2:06	70	"	"	37.5	Ethyl Alc. 1/5%
2:07	58	"	"	"	"
2:08					Pure
2:10	73	"	"	"	"
2:12	101	"	"	"	"
2:14	124.5	"	"	38	"
2:16	134	"	"	"	"
2:18	135	"	"	"	"
2:20					Ethyl Alc. 1/5%
2:21	110	"	"	"	"
2:23	91.5	"	"	"	"
2:25	83	"	"	38.5	"
2:27	70.5	"	"	"	"

Time	Amount pumped in 60 sec.			Temp.	Kind of blood.
2:29	61 C.C.	"		38.5°	Pure
2:32	104	"	"	"	"
2:35	130.5	"	"	"	"
2:37	133	"	"	39	"
2:39	134.5	"	"	"	"
2:42					Ethyl Alc. 1/5%
2:43	105	"	"	"	"
2:44	75	"	"	"	"
2:45					Pure
2:47	98	"	"	39.5	"
2:49	123	"	"	"	"
2:51	132	"	"	"	"
2:53	134	"	"	"	"
2:54					Ethyl Alc. 1/5%
2:55	100	"	"	"	"
2:56	68 " " Heart beat irregular			40	"
2:57			"	"	Pure
2:58	72 C.C. Kymograph tracing feeble.		"	"	"
2:59	64 C.C.	"	"	"	"
3:	40 " Heart dying		"	40.5	"

25

RESULTS:

At 35° a fall from 135.5 C.C. to 130 C.C. or reduction of 5.5 C.C
" 36° " " " 137.5 " " 123 " " " " 14.5 "
" 37° " " " 136 " " 115 " " " " 21. "
" 38° " " " 135 " " 110 " " " " 25. "
" 30° " " " 134.5 " " 103 " " " " 31.5 "
" 39.5 " " " 134 " " 100 " " " " 34. "

Average taken from the measurements at the temperature of 36° 37° 38° is 20.16 C.C. reduction at 37° in amount pumped around in 30 seconds.

EXPERIMENTS, May 8th, 1886. Ethyl Alcohol 1/5% in defibrinated calf's blood. Dog under Morphine and Curare. Isolation completed 2 P.M. Heart under experimentation 70 minutes.

------------------:---------------

Time	Amount pumped in 60 sec.	Temp.	Kind of blood.
2:05	164 C.C. "	35.5°	Pure
2:06	163 " "	"	"
2:07	164 " "	"	"
2:08	165 " "	36	"
2:09		"	Ethyl Alc. 1/5%
2:10	153 " "	"	"
2:11	142 " "	"	"
2:12	129 " "	"	"
2:13	110 " "	36.5	"
2:14	98 " "	"	"
2:15			Pure
2:16	113 " "	"	"
2:17	135 " "	"	"
2:18	156 " "	37	"
2:19	162 " "	"	"
2:20	164 " "	"	"
2:21			Ethyl Alc. 1/5%
2:22	148 " "	"	"
2:23	130 " "	"	"

Time	Amount pumped in 60 sec.	Temp.	Kind of blood.
2:26	94 C.C. "	37°	Ethyl Alc. 1/5%
2:27			Pure
2:28	100 " "	38	"
2:30	138 " "	"	"
2:34	102 " "	"	"
2:36	103 " "	"	"
2:38			Ethyl Alc. 1/5%
2:39	148.5 " "	"	"
2:40	124 " "	"	"
2:41	103 " "	39	"
2:42	82 " "	"	"
2:43	64 " "	"	"
2:44			Pure
2:46	88 " "	"	"
2:49	138 " "	39.5	"
2:51	138.5 " "	"	"
2:52	102 " "	"	"
2:54	102 " "	"	"
2:58	102.5 " "	"	"
2:59			Ethyl Alc. 1/5%
3:	138 " "	40	"
3:01	114 " "	"	"
3:02	90 " "	"	"

Time	Amount pumped in 30 sec.	Temp.	Kind of Blood.
3:03	55 C.C. " Heart irregular.	40°	Ethyl Alc. 1/5%
3:04		"	Pure
3:06	54 C.C. "	"	"
3:08	65 " " Irregular	"	"
3:09	" 48 " "	"	"
3:10	42 " " Kymograph tracing feeble.	"	"

Heart now pumped less and less and finally stopped at 3:15

RESULTS:

At 36° a fall from 165 C.C. to 153 C.C. or red. of 12 C.C.) In
" 37° " " " 164 " " 148 " " " " 16 ") amount
) pumped
" 38° " " " 163 " "143.5 " " " " 19.5") around
) in 30
" 40° " " " 162.5 " " 138 " " " " 24.5") sec.

Average taken from the measurement at the first three temperatures - 15.83 C.C. reduction at 37°

EXPERIMENT May 7th, 1886, Ethyl Alcohol 1/5% in defibrinated calf's blood.   Dog under Morphine and Curare.   Heart isolated 1:25 P.M. good for 1,1/2 hour.

------------:-------------

Time	Amount pumped in 30 sec.	Temp.	Kind of blood.
1:30	143 C.C. "	36°	Pure
1:32	150 " "	"	"
1:33	149 " "	"	"
1:35	150 " "	"	"
1:35 45 sec.			Alcohol 1/5%
1:37	141:5 " "	"	"
1:39	130 " "	36.5	"
1:40	122 " "	"	"
1:42	101 " "	"	"
1:43	96 " "	"	"
1:45	85 " "	"	"
1:46			Pure
1:48	94.5 " "	37	"
1:49	103.5 " "	"	"
1:50	132 " "	"	"
1:52	142 " "	"	"
1:53	148 " "	"	"
1:54			Ethyl Alc. 1/5%

Time	Amount pumped in 30 seconds		Temp.	Kind of blood
1:55	130 C.C.	"	37	Ethyl Alc. 1/5%
1:56	112 "	"	"	"
1:57	95 "	"	37.5	"
1:58	78.5 "	"	"	"
2:				Pure
2:01	92 "	"	"	"
2:02	110 "	"	"	"
2:03	128.5 "	"	"	"
2:04	142 "	"	"	"
2:06	144.5 "	"	"	"
2:08	146 "	"	38	"
2:10	146 "	"	"	"
2:11				Ethyl Alc. 1/5%
2:12	125.5 "	"	"	"
2:13	102 "	"	"	" "
2:14	80 "	"	"	"
2:15	60.5 "	"	38.5	"
2:16				Pure
2:17	85 "	"	"	"
2:19	104.5 "	"	"	"
2:21	128 "	"	"	"
2:25	140 "	"	"	"
2:28	144.5 "	"	39	"
2:29				Ethyl Alc. 1/5%

Time	Amount pumped in 30 seconds	Temp.	Kind of blood.
2:30	120 C.C. "	39°	Ethyl alc. 1/5%
2:33	94 " "	"	"
2:35	80 " "	"	"
2:38	58 " "	"	"
2:40			Pure
2:42	69 " "	39.5	"
2:44	96.5 " "	"	"
2:46	124 " "	"	"
2:48	138.5 " "	"	"
2:49	140 " "	"	"
2:51	142 " "	40	"
2:53			Ethyl Alc. 1/5%
2:54	111 " "	"	"
2:56	89 " " irregular pulse.	"	"
58	Did not recover under pure blood.		Pure

RESULTS.

At 36° a fall from 150 C.C. to 141.5 C.C. or red. of 8.5 C.C) In
" 37° " " " 143 " " 130 " " " " 18 ")amount
" 38' " " " 146 " " 125.5 " " " " 20.5 ")pumped
" 39' " " " 144.5 " " 120 " " " " 24.5 ")around
" 40° " " " 142 " " 111 " " " " 31 ")in 30
)seconds.

The average reduction calculated from the first three measurements above is 15.5 C.C. at 37° in 30 seconds.

The greater effect of the alcohol at higher temperature is very noticeable in this experiment the successive reductions being 8.5 C.C. 13. C.C. 20.5 C.C. 24.5 C.C. and 31. C.C. at 36° 37° 38° 39° and 40° respectively.

EXPERIMENT, May 10th, 1886. Ethyl Alcohol 1/5% in defibrinated calf's blood. Dog under Morphine and Curare,. Heart isolated 2:24 P.M. remaining good 1,1/2 hour.

------------------+------------------

Time	Amount pumped in 30 sec.	Temp.	Kind of blood.
2:30	128 C.C. "	35	Pure
2:32	130 " "	"	"
2:34	130.5 " "	"	"
2:36			Ethyl Alc. 1/5%
2:37	120 " "	"	"
2:38	112 " "	"	"
2:39	101 " "	35.5	"
2:41	80 " "	"	"
2:42			Pure
2:44	103:5 " "	"	"
2:46	125 " "	"	"
2:48	130 " "	36	"
2:49	129.5" "	"	"
2:50			Ethyl Alc. 1/5%
2:51	118 " "	"	"
2:53	90.5 " "	"	"
2:55	68 " "	"	"
2:57			Pure

Time	Amount pumped in 30 sec.			Temp.	Kind of blood.
2:59	84.5 C.C.	"		36.5	Pure
3:	109.5	"	"	"	"
3:02	129	"	"	"	"
3:04	129.5	"	"	"	"
3:06	129.5	"	"	37	"
3:08					Ethyl Alc. 1/5%
3:10	110	"	"	"	"
3:12	91.5	"	"	"	"
3:14	69	"	"	"	"
3:15				37.5	Pure
3:17	75	"	"	"	"
3:19	96	"	"	"	"
3:22	128	"	"	"	"
3:24	129	"	"	"	"
3:26	"	"	"	38	"
3:28					Ethyl Alco. 1/5%
3:29	105	"	"	"	"
3:31	78	"	"	"	"
3:33 Heart murmur	49	"	"	"	"
3:34					Pure
3:36	74.5	"	"	38.5	"
3:38	112	"	"	"	"
3:40	126	"	"	"	"

Time	Amount pumped in 30 sec.	Temp.	Kind of blood.
3:42	128.5 C.C. "	38.5²	Pure
3:44	128 " "	39	"
3:46			Ethyl Alcohol 1/5%
3:47	98.5 " "	"	"
3:49	69 " "	"	"
3:51	34 " "	39.5	"
3:52			Pure
3:53	71 " "	"	"
3:55	102.5 " "	"	"
3:57	124.5 " "	40	"
3:58	125. " "	"	"
3:59			Ethyl Alc. 1/5%
4	91 " " irregular	"	"
4:01	30 Kymograph)	"	
4:02	tracing) faint)		Pure
			"
4:03	heart dying.		

RESULT:

At 35 a fall from 130.5 C.C. to 120 C.C. or red. of 10.5 C.C.)   In
)
" 36 " "   "   129.5 " " 118 " "   "   " 11.5 ")amount
)
" 37 " "   "   129.5 " " 110 " "   "   " 19.5 ")pumped

30

At 38   a fall from 129 C.C. to 105 C.C. or red. of 24 C.C.)around
" 39   "   "   "   128 "   " 93.5 "   "   "   " 29.5 ")in 30
" 40   "   "   "   125 "   " 91   "   "   "   " 34   ")seconds

Average taken from the measurements at 36 37 and 38 is 18.33 C.C. reduction at 37° in amount pumped around in 30 seconds. Lungs well preserved, no ecchymoses noticeable on outside of heart. For some reason which cannot yet be found out, a murmur (regurgitant) is heard over the left side of the heart only, it is most likely due to an insufficiency of the valves of the left side of the heart both of the Mitral and semilunar valves at the orifice of the aorta.

Average result of Four Experiments with Ethyl Alcohol 1/5% on defibrinated calf's blood.

May 4th, 1886, 20.16 C.C. at 37°)
)   Reduction in amount
 "   6th, "   15.83 "   "   "   )
)   pumped around
 "   7th, "   15.5   "   "   "   )
)   in 30 seconds.
 "   10th, " -18.33 "   "   "   )
            ----
          4)69.82(17.45

Average Total of 4 Experiments with 1/5% Ethyl Alcohol 17.45 C.C. reduction at 37° in amount pumped around in 30 seconds.

EXPERIMENT No II, May 20th, 1886. Propyl. Alcohol 1/5% in defibrinated calf's blood.   Isolation completed 3:18 P.M.   Heart under experimentation for 40 min.

Time	Amount pumped in 30 minutes	Temp.	Kind of blood
3:20	168 C.C. "	35.5°	Pure
3:21	170 " "	"	"
3:22	170 " "	"	"
3:23 30 sec.		36	Propyl Alc. 1/5%
3:24 15 "	105 " "	"	"
3:25	74 " "	"	"
3:26	42.5 " "	"	"
3:26 30 "		"	Pure
3:27 30 "	72 " "	36.5	"
3:28 15 "	100 " "	"	"
3:29	134.5 " "	"	"
3:30	164 " "	"	"
3:30 45 "	166 " "	"	"
3:31 30 "	168 " "	"	"
3:32 30 "	168 " "	37	"
3:33		"	Propyl Alc. 1/5%
3:33 50 "	85 " "	"	"
3:34 45 "	50.5 " "	"	"

Time	Amount pumped in 30 sec.			Temp.	Kind of blood.
3:35 30 sec.	25.5 C.C.	"		37°	Propyl. Alc. 1/5%
3:36				"	Pure
3:37	60.5	"	"	"	"
3:38	82	"	"	"	"
3:39	120	"	"	37.5	"
3:40	143	"	"	"	"
3:41	161	"	"	"	"
3:42	163	"	"	"	"
3:43	164	"	"	"	"
3:43 30 "			"	38	Propyl. Alc. 1/5%
3:44 20 "	75.5	"	"	"	"
3:45 38 "	45.5	"	"	"	"
3:46 30 "	25	"	"	"	"
3:47				"	Pure
3:48	49.5	"	"	"	"
3:49	78	"	"	38.5	"
3:50	110	"	"	"	"
3:51	160	"	"	"	"
3:51 45 "	159	"	"	"	"
3:52 15 "	160	"	"	"	"
3:53				39	Propyl. Alc. 1/5%
3:53 48 "	60.5	"	"	"	"
3:54 30 "	20	"	"	"	"

Time	Amount pumped in 30 sec.	Temp.	Kind of blood.
3:55	stopped "	39°	Propyl. Alc. 1/5%
3:55 30 sec.	"	"	Pure
3:56 15 "	30.5 C.C. "	39.5	"
3:57	58 " "	"	"
3:58	64 " "	"	"
3:59	91 " "	"	"
3:59 45 "	108 " "	"	"
4: 30 "	145 " "	40	"
4:01	148 " "	"	"
4:02	149 " "	"	"
4:03	150 " "	"	"
4:03 30 "			Propyl. Alc. 1/5%
4:04 15 "	30 " "	"	"
4:05	stopped"	"	"
4:06		40.5	Pure
Heart did not recover under pure blood		"	"

RESULT:

At 36 a fall from 170 C.C. to 105 C.C.or red. of 65 C.C.) In
" 37° " " " 168 " " 85 " " " " 83 ")amount
" 38° " " " 164 " " 75.5" " " " 88 ")pumped
)around
" 39° " " " 160 " " 60.5" " " " 99.5")in 30
)seconds
" 40° " " " 150 " " 30 " " " " 120 ")

40

The average reduction calculated from the first three measurements is 73.83 C.C. in 30 seconds at 37°

RESULTS of four Experiments with 1/5% Propyl. Alcohol

1. May 18th, 1886, N. 1   79.   C.C. reduction at 37 in 30 sec.
2. "   "   " " II 80.83 "   "   "   "   " " "
3. " 20th,   " " I   80.16 "   "   "   "   " " "
4. "   "   " " II 78.83 "   "   "   "   " " "
                       ----
              4)318.82(79.705

Total average reduction effected by 1/5% of Propyl. Alcohol in defibrinated calf's blood. (result of four experiments-
    = 79.705 C.C. at 37°
in amount pumped around by the isolated dog heart in 30 seconds.

EXPERIMENT N. II, May 18th, 1886. Propyl. Alcohol 1/5% in defibrinated dog's blood. Isolation completed 5:15 P.M. Heart under experimentation 45 min.

Time	Amount pumped in 30 sec.	Temp.	Kind of blood.
5:22	190 C.C. "	35°	Pure
5:23	190.5 " "	"	"
5:24	190.5 " "	35.5	"
5:25	190 " "	"	"
5:25 45 sec.	191 " "	"	"
5:26		36	C_2H_6O 1/5%
5:26 45 "	125 "	"	"
5:27 30 "	101 " "	"	"
5:28 15 "	84.5 " "	"	"
5:29	60 " "	36.5	"
5:29 30 "		"	Pure
5:30 15 "	102 " "	"	"
5:31	142.5 " "	"	"
5:32	169 " "	"	"
5:33	184 " "	37	"
5:34	185 " "	"	"
5:34 30 "		"	C_2H_6O 1/5%
5:35 15 "	100.5 " "	"	"
5:36	80 " "	37.5	"

42

Time	Amount pumped in 30 sec.	Temp.	Kind of blood.
5:37	40.5 C.C. "	37.5	C H O 1/5%
5:37 30 sec.		"	Pure
5:38 20 "	49 " "	"	"
5:39	86 " "	"	"
5:39 45 "	145 " "	38	"
5:40 40 "	168 " "	"	"
5:41 30 "	181 " "	"	"
5:42	182 " "	"	"
5:42 30 "	" "	38	C H O 1/5%
5:43 15	90 " "	"	Propyl. Alc. 1/5%
5:44	60 " "	"	"
5:45	32 " "	"	"
5:45 30 "		"	Pure
5:46 15 "	50 " "	38.5	"
5:47	84.5 " "	"	"
5:48	120.5 " "	"	"
5:49	164 " "	"	"
5:50	173 " "	39	"
5:51	180 " "	"	"
5:52 40 sec.	179 " "	"	"
5:53		"	Propyl Alc. 1/5%
5:54	80 " "	"	"
5:55	50 " "	"	"

Time	Amount pumped in 30 sec.	Temp.	Kind of blood.
5:56	20 C.C. "	39.5	Propyl Alc. 1/5%
5:56 30 sec.		"	Pure
5:57 40 "	40.5 " "	"	"
5:58 30 "	80 " "	"	"
5:59	110.5 " "	"	"
6:	134 " "	40	"
6:01	161.5 " "	"	"
6:01 35 "	162.5 " "	"	"
6:02 15 "	162 " "	"	"
6:03		"	Propyl. Alc. 1/5%
6:03 50 "	38 " "	"	"
6:04 10 "	stopped "	40.5	"
6:04 45 "		"	Pure
6:05 45 "	10 C.C. "	"	
6:07	Heart irregular did not recover under pure blood.		

RESULTS: Effects of 1/5% of Propyl Alcohol on the isolated dog heart are as follows:

After the last measurement at 40° the heart did not recover, hence it is not taken into account into making out the average reduction which was determined from the measurements at 36° 37° and 38° and is 80.83 C.C. reduction in amount pumped around in 30 seconds at 37°.

EXPERIMENT, V. I, May 18th, 1886. Propyl Alcohol 1/5% C_3H_8O.
Dog under Morphine and Curare. Defibrinated Calf's blood used.
Heart isolated 3:50. Remained good for 40 minutes.

---------------;--------------

Time	Amount pumped in 30 sec.	Temp.	Kind of blood.
3:55	168 C.C. "	36°	Pure
3:56	168.5 " "	"	"
3:57	170 " "	"	"
3:57 40 sec.	170.5 " "	"	"
3:58			Propyl 1/5%
3:58 45 "	101 " "	"	"
3:59 30 "	80 " "	"	"
4: 10 "	50 " "	"	"
4:01			Pure
4:01 45 "	68 " "	"	"
4:02 20 "	85 " "	"	"
4:03	109.5 " "	"	"
4:04	160 " "	36.5	"
4:04 30 "	165 " "	"	"
4:05 15 "	166 " "	"	"
4:06	166 " "	37	"
4:06 30 "		"	Propyl.
4:07 15 "	86.5 " "	"	"

Time	Amount pumped in 30 sec.		Temp.	Kind of blood.
4:08	50 C.C.	"	37°	Propyl. 1/5%
4:08 30 sec.	25.5 "	"	"	"
4:09		"		Pure
4:09 45	48 "	"	37.5	"
4:10 30 "	69 "	"	"	"
4:11	102.5 "	"	"	"
4:12	124 "	"	"	"
4:13	149 "	"	"	"
4:13 30 "	156 "	"	"	"
4:14 15 "	162 "	"	38	"
4:15	164 "	"	"	"
4:15 30 "	165 "	"	"	"
4:16		"	"	Propyl 1/5%
4:16 48 "	76 "	"	"	"
4:15 30 "	49 "	"	"	"
4:16	30 "	"	"	"
4:16 35 "				Pure
4:17 30 "	48 "	"	"	"
4:18	75 "	"	"	"
4:18 30 "	118 "	"	38.5	"
4:19 30 "	132 "	"	"	"
4:20 15 "	151.5 "	"	"	"
4:21	159 "	"	"	"

Time	Amount pumped in 30 sec.		Temp.	Kind of Blood.
4:21 45 sec.	160 C.C.	"	38.5°	"
4:22 30 "	161 "	"	39	"
4:23 15 "	160 "	"	"	"
4:24				Propyl 1/5%
4:24 45 "	62 "	"	"	"
4:25 30 "	30 "	"	"	"
4:26				Pure
4:27	58 "	"	"	"
4:27 45 "	76 "	"	"	"
4:28 30 "	120 "	"	39.5	"
4:29 30 "	130 "	"	"	"
4:30 15 "	131 "	"	"	"
4:31	130 "	"	"	"
4:31 30 "		"	"	Propyl 1/5%
4:32	few drops about 2 C.C. then stopped		40	"
4:33	Heart very irregular and weak, Kymograph tracing very faint- pure blood turned on, but did not recover it.			Pure

It is probable that the temperature of $40°$ was too high the last time the propyl alcohol was turned on, otherwise the heart would not have ceased pumping around so suddenly.

This reduction at 4 o'clock 32 min. however allows us to judge of the effect of 1/5% propyl alcoholized blood at $40°$ which according to the measurement must be about 128 C.C. in 30 seconds.

The heart was very much enlarged, all over it were scattered dark red ecchymoses varying from size of a pin head to that of a pea. A strange observation was made during this experiment i.e. the left ventricle ceased contracting and the right ventricle continued so for some minutes after 4:32 min. The lungs were very oedematous, and also covered with ecchymoses.

RESULTS of 1/5% Propyl Alc. blood on isolated dog's heart May 18th, 1886.

At $36°$ a fall from 170.5 C.C. to 101 C.C. or red. of 68.5 C.C. in 30 sec
" $37°$ " " " 186 " " 86.5 " " " " 79.5 " " " "
" $38°$ " " " 165 " " 76 " " " " 89 " " " "
" $39°$ " " " 160 " " 62 " " " " 98 " " " "

Average from measurements at $36°$ $37°$ and $38°$

Reduction 79 C.C. in 30 sec. at $37°$.

EXPERIMENT N.I, May 20th, 1886 with 1/5% Propyl Alcohol in C_2H_6O defibrinated calf's blood. Dog under Morphine and Curare. Heart isolated 1:30 P.M. Heart remained good for

----

Time	Amount pumped in 30 sec.	Temp.	Kind of blood.
1:34	178 C.C. "	35.5°	Pure
1:35	179 " "	"	"
1:36	180 " "	"	"
1:37	180 " "	"	"
1:37 30 sec.	"	36	C_2H_6O 1/5%
1:38 15 "	116 " "	"	"
1:39	84 " "	"	"
1:39 30 "	50 " "	"	"
1:40		36.5	Pure
1:40 45 "	90 " "	"	"
1:41 30 "	120.5 " "	"	"
1:42 45 "	146.5 " "	"	"
1:43 30 "	173 " "	"	"
1:44	176 " "	37	"
1:44 45 "	178 " "	"	"
1:45 30 "	177 " "	"	"
1:46			C_2H_6O 1/5%
1:46 45 "	92 " "	"	"

Time	Amount pumped in 30 sec.			Temp.	Kind of blood.
1:47 30 sec.	30.5	"	"	37°	C_3H_7O 1-5%
1:48 30 "	32	"	"	"	"
1:49			"		Pure
1:49 45 "	53	"	"	37.5	"
1:50 30 "	84.5	"	"	"	"
1:51 15 "	120	"	"	"	"
1:52	138	"	"	"	"
1:52 45 "	164	"	"	"	"
1:53 30 "	174	"	"	"	"
1:54	175	"	"	38	"
1:54 45 "	174	"	"	"	"
1:56				"	C_3H_7O 1/5%
1:56 40 "	82.5	"	"	"	"
1:57 15	50	"	"	"	"
1:58	21.5	"	"	38.5	"
1:58 45 "	stopped			"	
1:59				"	Pure
2:	56	"	"	"	"
2:45	83.5	"	"	"	"
2:01 30 "	110	"	"	"	"
2:02 45 "	136	"	"	"	"
2:03	168	"	"	"	"
2:03 45 "	169	"	"	39	"
2:04 30 "	168	"	"	"	"

Time	Amount pumped in 30 sec.	Temp.	Kind of blood.
2:05			C_2H_6O 1/5%
2:05 40 sec.	68.5 C.C. "	39°	"
2:06 30 "	20 " "	."	"
2:07	stopped "	"	"
2:07 30 "		39.5	Pure
2:08	Began again but very irregular--- did not recover perfectly under pure blood.		

RESULTS of experiment N.I May 20th, 1886.

In experiment N. I of May 20th, 1886 the following effects of 1/5% Propyl Alcohol are noted according to this table. At 36° a fall from 180 C.C. to 116 C.C. or red. of 64 C.C. in 30 sec.

" 37° " " " 177 " " 92 " " " " 85 " " " "

" 38° " " " 174 " " 82.5" " " " 91.5" " " "

" 39° " " " 168 " " 68.5" " " " 99.5" " " "

After Propyl Alcohol was turned on at 39° the heart became arythmical and pumped around so weakly that experiment was discontinued.

Average taken from the first three measurements.
Reduction 80.16 C.C. at 37° in 30 sec.

Heart stopped in Diastole very much swollen and protruding from a rent in the pericardium.

Heart and lungs covered all over with ecchymoses, which were apparent also on cutting into these organs.

EXPERIMENT .I May 3th, 1888. Isobutyl Alcohol C_4H 1/5% in defibrinated calf's blood. Dog under Morphine and Curare. Heart isolated 2.8 P.M. remained good for 30 min.

Time	Amount pumped in 30 sec.	Temp.	Kind of blood.
2:10	196 C.C. "	35.5	Pure
2:10 30 sec.	197 " "	"	"
2:11	198 " "	"	"
2:11 30 "	197 " "	36	"
2:12			Butyl Alc. 1/5%
2:12 45 "	45 " "	"	"
2:13 30 "	5 " "	"	"
2:13 40 "	stopped "	"	"
2:14			Pure
2:14 45 "	24.5" "	36.5	"
2:15 30 "	73 " "	"	"
2:16 15 "	122 " "	"	"
2:17 10 "	153.5" "	"	"
2:18	179 " "	"	"
2:18 40 "	195 " "	"	"
2:19 15 "	194 " "	37	"
2:19 45			Butyl Alc. 1/5%
2:20 30 "	32.5" "	"	"
2:21	3 " "	"	"

Time	Amount pumped in 30 sec.			Temp.	Kind of blood.
2:21 10 sec.	stopped		"	37	Butyl Alc. 1/5%
2:21 45	"			37.5	Pure
2:22 30	"	41.5 C.C.	"	"	"
2:23 30	"	85	"	"	"	"
2:24 30	"	140.5	"	"	"
2:25 30	"	136	"	"	"	"
2:26 30	"	194.5	"	"	38	"
2:27 15	"	192.5	"	"	"	"
2:27 45	"				Butyl Alc. 1/5%
2:28 30	"	21	"	"	"	"
2:29		stopped	"	"	"
2:29 40	"			"	Pure
2:30 30	"	35 C.C.	"	"	"
2:31 30		68.5	"	"	38.5	"
2:32 30	"	97.5	"	"	"	"
2:33 30	"	129	"	"	"	"
2:35		136	"	"	"	"
2:36		136	"	"	39	"
2:36 15	"			"	Butyl Alc. 1/5%
2:37	a few drops about 2 C.C.		"	"	
2:37 10	then stopped		"	"	

Time	Amount pumped in 30 sec.	Temp.	Kind of blood.
2:37 40 sec.			Pure
2:38	heart did not begin to pump around again under pure blood.		

RESULTS of 1/5% solution of primary Isobutyl Alcohol in defibrinated calf's blood on isolated dog heart.

At $36°$ a fall from 197 C.C. to 45 C.C. or red. of 152 C.C. in 30 sec.
" $37°$ " " " 194 " "32.5 " " " "161.5 " " " "
" $38°$ " " " 192.5 " " 21 " " " "171.5 " " " "
" $39°$ " " " 136 " " 2 " " " "136 " " " "

After the last measurement at $39°$ the heart did not recover.

Average 161.66 C.C. at $37°$ in 30" (taken from first three measurements). In the experiments with Butyl and Amyl Alcohols measurements were taken as often as possible.

Four large graduated glasses were placed in a row next to the outflow tube which was bent like a hook. As soon as thirty seconds were counted during which blood was measured into one graduate, with the 31st second, the glass, (bent end) of the outflow tube was hooked over another empty graduate while the first was noted and emptied back into supply bottle and so on with the 3rd and 4th graduate. By the time the fourth was used, the first graduate was again ready for measuring.

May 24th, 1836, I Experiment.

Heart and lungs covered with ecchymoses the size of a large pea. Lungs very oedematous.

Left ventricle at end of experiment ceased beating before the right.

EXPERIMENT. No. II.   May 24th, 1886.   Isobutyl Alcohol 1/5% in defibrinated calf's blood.   Heart isolated 3:15 P.M. and remained under experimentation for 35 minutes.

Time	Amount pumped in 30 sec.	Temp.	Kind of blood.
3:18	189 C.C. "	35.°	Pure
3:18 45 sec.	191.5 " "	"	"
3:19 30 "	193.5 " "	"	"
3:20 20 "	195 " "	35.5	"
3:20 45 "		"	Butyl Alc. 1/5%
3:21 30 "	45 " "	"	"
3:22	15 " "	"	"
3:22 20 "	Stopped "	36	"
3:22 45 "		"	Pure
3:23 20 "	40 C.C.	"	"
3:24	92 " "	"	"
3:25	149 " "	36.5	"
3:26	190 " "	"	"
3:26 30 "	193 " "	"	"
3:27	194 " "	"	"
3:27 30 "		37	Butyl Alc. 1/5%
3:28 15 "	34 " "	"	"
3:28 45 "	2 " "	"	"
3:29	stopped "	"	"
3:29 15 "		"	Pure

Time	Amount pumped in 30 sec.			Temp.	Kind of blood.
3:29 50 sec.	48 C.C.		"	37.5	Pure
3:30		85.5	"	"	"
3:30 40 "	124		"	"	"
3:31 30 "	142.5		"	"	"
3:32 15 "	168		"	"	"
3:33		186	"	"	"
3:34		189.5	"	38	"
3:34 40 "	190		"	"	"
3:35 15 "	190		"	38	"
3:36		189.5	"	38.	"
3:36 45 "	189		"	38.5	"
3:37 30 "	190		"	"	"
3:38 10 "				"	Butyl Alc. 1/5%
3:38 20 "	18		"	"	"
3:38 45 "	stopped		"	"	"
3:39 30 "				"	Pure
3:40 30 "	20 C.C.		"	39	"
3:41 30 "	40		"	"	"
3:42 15 "	89		"	"	"
3:42 48 "	121.5		"	"	"
3:43 20 "	169		"	"	"
3:44		182.5	"	"	"

Time	Amount pumped in 30 sec.			Temp.	Kind of blood.
3:45	185 C.C.	"		39.5°	Pure
3:45 30 sec.	186	"	"	"	"
3:46				"	Butyl Alc. 1/5%
3:46 10	"	about 1	"	"	"
3:46 38	"	stopped	"	"	"
3:47 15	"			40	Pure
3:48		15	"	"	"
3:48 30	"	20	"	"	"
3:49		12	"	"	"

Kymograph tracing feeble--- heart very irregular.

RESULTS of action of 1/5% Isobutyl Alcohol on isolated heart.

At 35.5° a fall from 195 C.C. to 45 C.C. or red of 150 C.C.) In
" 37° " " " 194 " " 34 " " " " 160 ") amount) pumped
" 38.5° " " " 190 " " 18 " " " " 172 ") around) in 30
" 39.5° " " " 186 " " 1 " " " " 185 ") seconds

Average reduction 160.66 C.C. in 30 sec. at 37° (taken from first three measurements).

EXPERIMENT May 25th, 1886. Isobutyl Alcohol 1/5 in defibrinated Dog's blood. Dog under Morphine and Curare. Heart isolated 2:24 P.M. Remained good for 32 minutes.

------------+------------

Time	Amount pumped in 30 sec.		Temp.	Kind of Blood.
2:26	184 C.C.	"	35.5	Pure
2:26 45 sec.	185 "	"	"	"
2:27 30 "	186 "	"	"	"
2:28 15 "	186.5"	"	36	"
2:28 45 "				Butyl Alc. 1/5%
2:29 30	82.5"	"	"	"
2:30	5 "	"	"	"
2:30 5 "	stopped		36.5	
2:30 35 "			"	Pure blood.
2:31 40	48 "	"	"	"
2:32 30	76.5"	"	"	"
2:33	98 "	"	"	"
2:34	134.5"	"	"	"
2:35	186.5"	"	37	"
2:36	184.5"	"	"	"
2:36 40 "	186 "	"	"	"
2:37				Butyl Alc. 1/5%
2:37 40 "	24 "	"	"	"
2:38 10 "	stopped	"	"	"

Time	Amount pumped in 30 sec.	Temp.	Kind of Blood.
2:38 40 sec.			Pure
2:39 30 "	36.5 C.C. "	37.5°	"
2:40	69 " "	"	"
2:41	93.5 " "	"	"
2:42	126 " "	"	"
2:43	172 " "	"	"
2:44	182 " "	"	"
2:44 45 "	184 " "	38	"
2:45 30	184	"	"
2:46		"	Butyl Alc. 1/5%
2:46 40 "	135 " "	"	"
2:46 45 "	stopped "	"	"
2:47			Pure
2:48 40 "	42 " "	"	"
2:49 30 "	84.5 " "	38.5	"
2:51	114 " "	"	"
2:52	136.5 " "	"	"
2:53	166 " "	"	"
2:54	180.5 " "	"	"
2:55	181.5 " "	39	"
2:55 40 "	183.5 " "	"	"
2:56 15 "	184 " "	"	"
2:56 30		"	Butyl Alc. 1/5%

Time	Amount pumped in 30 sec.	Tem.	Kind of Blood.
2:57	Heart stopped 30 seconds after the drug was turned on.	36.5°	Butyl Alc. 1/5?
2:57 30 sec.		"	Pure
2:58	began pumping around again but very irregular, heart did not make a perfect recovery under pure blood.		

RESULTS:

At 36 a fall from 186.5 C.C. to 32.5 C.C. or Red. of 154 C.C.)
" 37 " " " 186 " " 24 " " " " 162 ") In
" 38 " " " 184 " " 13.5 " " " " 170.5 ") 30
" 39 " " " 184 " " 0 " " " " 184) seconds

Average drawn from first three measurements.

162.166 C.C. in 30 seconds at 37°.

May 26th, 1886.
The dog used to-day and those used yesterday did not differ in weight more than 100 Grams, still the results of the experiments differ considerably.

The older the dog the better does the heart stand the experiment, Young

dogs of large size will not stand the operation as well as old dogs of smaller size.

In the experiment on May 25th, 1886 the lungs became very oedematous and covered with large ecchymoses, internal hemorrhages had taken place into the alveoli of the lung, clogging up the capillary bronchial tubes with blood, mucus and froth.

Heart also covered with ecchymoses.
Results of three experiments with 1/5% Butyl(iso) alcohol
May 24th, N.I 161.66 C.C. reduction at 37° in 30 sec.
  " "   N.II 160.66 "   "   " " " " "
 " 25   162.166 "   "   " " " " "
        ----
        3)484.436(161.495 C.C. reduction at 37° in 30 sec.

EXPERIMENT May 28th, 1886. Isobutyl Alcohol 1/10% in defibrinated Calf's Blood. Dog under Morphine and Curare. Heart isolated at 2:15 P.M. Remained good for 40 min..

Time	Amount pumped in 30 sec.	Temp.	Kind of Blood.
2:18	174 C.C. "	35°	Pure
2:18 45 sec.	175 " "	"	"
2:19 15 "	176 " "	"	"
2:20	173 " "	"	"
2:20 30	178.5 " "	"	"
2:21	"		Butyl Alc. 1/10%
2:21 40 "	119 " "	"	"
2:22 30 "	89 " "	35.5	"
2:23	42.5 " "	"	"
2:23 30 "		"	Pure
2:24 15 "	90 " "	"	"
2:25	160.5 "	36	"
2:26	171 " "	"	"
2:27	174.5 " "	"	"
2:28	177 " "	"	"
2:29	177 " "	"	"
2:29 30 "			Butyl Alc. 1/10%
2:30 10 "	102 " "	"	"
2:31	64.5 " "	"	"

Time	Amount pumped in 30 sec.			Temp.	Kind of blood.
2:31 40 sec.	35 C.C.		"	36.5	Butyl Alc. 1/10%
2:32 20	"				Pure
2:33	55	"	"	"	"
2:34	102.5	"	"	"	"
2:35	143	"	"	"	"
2:36	170	"	"	"	"
2:37	174	"	"	37	"
2:38	175	"	"	"	"
2:39	175.5	"	"	"	"
2:39 30	"			"	tyl Alc. 1/10%
2:40 15	"	95	"	"	"
2:41	40	"	"	"	"
2:41 30	"				Pure
2:42 40	"	84.5	"	"	37.5	"
2:43 30	120	"	"	"	"
2:44	141	"	"	"	"
2:45	168.5	"	"	"	"
2:46	173	"	"	"	"
2:47	174	"	"	38	"
2:47 30	174	"	"	"	"
2:48				"	Butyl Alc.
2:48 45	"	89.5	"	"	"	"
2:49 20	"	40.5	"	"	"	"

Time	Amount pumped in 30 sec.		Temp.	Kind of blood.
2:50				Pure
2:51	74 C.C.	"	38.5	"
2:52	110 "	"	"	"
2:53	156 "	"	"	"
2:54	164 "	"	"	"
2:55	170 "	"	"	"
2:56	169.5 "	"	39	"
2:56 30 sec.		"	"	Butyl Alc. 1/10%
2:57 10 "	68 "	"	"	"
2:58	heart stopped	"	"	"
2:58.15 sec. Kymograph out order				Pure
2:58	60 C.C.	"	"	"
2:59 30	98.5 "	"	"	"
3:	110 "	"	40	"
3:04	An accident to the Kymograph forced interruption to experiment.			

Result of Experiment May 26th, 1886, Isobutyl Alcohol 1/10%

At 35° a fall from	178.5	C.C. to	199.C.C.	or red. of	59.5 C.C.	)					
" 36° " "	"	177	"	" 102	"	"	"	" 75	"	)	In
" 37 " "	"	175.5	"	" 95	"	"	"	" 80.5	"	)	30
" 38 " "	"	174	"	" 89.5	"	"	"	" 84.5		)	seconds
" 39 " "	"	169.5	"	" 68	"	"	"	" 101.5		)	

These are effects of 1/10% Isobutyl Alcohol, the average reduction taken from the three measurements at 36°, 37° 38° is 80 C.C. at 37° in 30".

If we could judge of the effect of 1,10% Isobutyl Alcohol in defibrinated calf's blood, by doubling this result we should have 160 C.C. at 37° in 30".

The animal used in the experiment on May 26th was of same size and apparently age, as the one used on the day before, as it was brought from a farm were it was raised with others of the same litter. This dog was 1,1/2 lb. heavier than the one used on the previous day (May 25th), still the average reduction in the amount pumped around at 37° under Isobutyl Alcohol 1/10 differs from that reduction observed on May 26th, 1886. The percentage on that day was 1/5 Isobutyl we would presume that the effect observed was twice as great as shown by 1/10% May 26th. But although the dog was lighter the effect was greater.

The average on May 26th, being   160.   C.C. at 37 in 30 sec.
"   "   " " 25,   "   162.166 "   " "   "   "
"   "   " " 24,M.I   "   161. 66 "   " "   "   "
"   "   " " 24,M.II "   160. 66 "   " "   "   "
                                 ----
                              4)644.486(161.121 C.C.

Average of four experiments with Isobutyl Alcohol 161.121C.C. reduction.

Average reduction in three experiments with 1/5% Isobutyl Alcohol 161.495 C.C. at 37° in 30 "

EXPERIMENT N.II   June 4th, 1880.   Amyl Alcohol 1/10% in defibrinated dog's blood.   Animal under Morphine and Curare.   Heart under experimentation for 54 min?

-------------------:---------------

Time	Amount pumped in 60 sec.			Temp.	Kind of Blood.
3:31	204 C.C.	"		39.5°	Pure
3:31 45	205	"	"	"	"
3:32 30 sec.	205.5	"	"	39	"
3:33				"	Amyl Alc. 1/10%
3:33 45	"	5	"	"	"
3:34 30	"			"	Pure
3:35	49	"	"	38.5	"
3:36	153	"	"	"	"
3:37	194	"	"	"	"
3:38	203	"	"	38	"
3:39	204	"	"	"	"
3:39 30	"			"	Amyl 1/10%
3:40 15	"	24.5	"	"	"
3:41					Pure
3:42	68	"	"	37.5	"
3:43	124.5	"	"	"	"
3:44	169	"	"	"	"
3:45	198.5	"	"	"	"
3:46	202	"	"	37	"

Time	Amount pumped in 30 sec.			Temp.	Kind of blood.
3:47	203 C.C.		"	37°	Pure
3:47 30 sec.					Amyl Alc. 1/10%
3:48 20	"	45	"	"	"
3:49					Pure
3:50		93.5 "	"	36.5	"
3:51		148	"	"	"
3:52		196.5 "	"	"	"
3:53		200	"	"	"
3:54		201	"	"	"
3:54 20	"			36	Amyl. Alc. 1/10%
3:55		82	"	"	"
3:55 45		5	"	"	"
3:56	heart ir-regular				
3:56 30	"				Pure
3:57		30	"	"	"
3:58		98.5 "	"	35	"
3:59		148	"	"	"
4:		190	"	"	"
4:01		192	"	"	"
4:01 30	"				Amyl Alc. 1/10%
4:02		70	"	"	"
4:02 45	"	4 heart irreg-ular.			

Time	Amount pumped in 30 sec.	Temp.	Kind of blood.
4:03	stopped		Amyl Alc. 1/10%
4:03	45 sec.		Pure
4:05	heart did not recover under pure blood.		

In this experiment the temperature was intentionally made high at the beginning and gradually lowered as the experiment progressed.

RESULTS, on isolated dog's heart of 1/10% Amyl Alcohol.

At 39 a fall from 205.5 C.C. to 5 C.C. or red. of 200.5C.C.) In
" 38   "   "   "   204   "   "24.5 "   "   "   " 179.5 ")Amount
" 37   "   "   "   203   "   "45   "   "   "   " 158   ")pumped
" 36   "   "   "   201   "   "62   "   "   "   " 139   ")in 30
" 35   "   "   192   "   "70   "   "   "   " 122   ")sec.

Notice as the temperature decreases the effect of the Alcohol decreases.

Average reduction 153.83 C.C. at 37 in amount pumped around in 30 sec.

EXPERIMENT, May 27th, 1886. Amyl Alcohol 1/10% in defibrinated dog's blood, Animal under Morphine and Curare. Isolation completed 2:25 P.M. Heart under experiment.

Time	Amount pumped in 30 sec.	Temp.	Kind of blood
2:28	208 C.C. "	35	Pure
2:29	207 " "	"	"
2:30	207.5 " "	"	"
2:31	208 " "	36	"
2:31 30 sec.		"	Amyl Alc. 1/10%
2:32 20 "	68 " "	"	"
2:33		"	Pure
2:34	94 " "	36.5	"
2:36	143 " "	"	"
2:38	192 " "	"	"
2:40	206 " "	"	"
2:41	207 " "	37	"
2:41 25 "		"	Amyl Alc. 1/10%
2:42	48 " "	"	"
2:42 30 "			Pure
2:44	91.5 " "	"	"
2:46	144 " "	37.5	"
2:48	193 " "	"	"
2:50	207 " "	"	"
2:50 30			Amyl Alc. 1/10%

Time	Amount pumped in 30 sec.			Temp.	Kind of blood.
2:51 20 sec.	26 C.C.		"	38°	Amyl Alc. 1/10%
2:52					Pure
2:54	89.5	"	"	"	"
2:56	129	"	"	"	"
2:58	193	"	"	38.5	"
2:59	206	"	"	"	"
3	205.5	"	"	"	"
20	"				Amyl Alc. 1/10%
3:04 10	"	5	"	"	39	"
3:04 35	"				Pure
3:05 40	"	61	"	"	"	"
3:06 40	"	79	"	"	"	"
3:09	139	"	"	"	"
3:10	198	"	"	39.5	"
3:11	197	"	"	"	"
3:12	197	"	"	"	"
3:12 20				"	Amyl Alc. 1/10%
3:13	stopped pumping.			40	"
3:13 30	heart very irregular				Pure
3:14	did not recover under pure blood.				

I

RESULTS:

1/10% of Amyloxyhydrate in this experiment produced the following reductions in the amounts pumped around by an isolated dog heart in 30 sec.

At 36° a fall from 208 C.C. to 68 C.C. or red. of 140 C.C.)
" 37° "   "   "   207   " " 48 " " "   " 159 "   )
" 38° "   "   "   207   " " 29 " " "   " 178 "   )
" 39° "   "   "   205.5" " " 5 " " "   " 200.5"   )

Average reduction 159 C.C. at 37°

This is the only experiment in which Amyl Alcohol was successfully turned on at 39°, its effect at that temperature seems to be about 200 C.C. reduction in the amount pumped around in 30 seconds.

EXPERIMENT N.I June 3rd, 1886. Amyl Alcohol 1/10% in defibrinated calf's blood. Dog under Morphine and Curare. Isolation completed 1:40 P.M. Heart lived 26 min. 30 sec.

Time	Amount pumped in 30 sec.	Temp.	Kind of blood.
1:45	193 C.C. "	35.5	Pure
1:46	194 " "	"	"
1:46 30 sec.	194 " "	"	"
1:47 15 "	193.5 " "	36	"
1:48		"	Amyl Alc. 1/10%
1:48 45 "	43 " "	"	"
1:49 15 "	stopped	36.5	"
1:49 45 "		"	Pure
1:50 30 "	83 " "	"	"
1:51	111 " "	"	"
1:52	182.5 " "	"	"
1:52 30 "	184 " "	"	"
1:53 15 "	184.5 " "	37	"
1:53 45 "		"	Amyl Alc. 1/10%
1:54 30 "	25 " "	"	"
1:55	stopped "	37.5	"
1:55 30		"	Pure
1:56 15 "	20 " "	"	"
1:57	94.5 " "	"	"
1:57 45	124 " "	"	"

Time	Amount pumped in 30 sec.				Temp.	Kind of blood.
1:58 30 sec.	181 C.C.		"		37.5°	Pure
1:59 15 "	182		"		"	"
2:	182.5		"		38	"
2: 30 "	182		"		"	"
2:01 15 "					"	Amyl Alc. 1/10%
2:02 30	" a few drops about 1 C.C.				"	"
2:02 40	stopped				"	"
2:03					"	Pure
2:03 45 "	10		"		38.5	"
2:04 30 "	24		"		"	"
2:05 15 "	49		"		"	"
2:06	93		"		"	"
2:06 35 "	120		"		"	"
2:07	124		"		39	"
2:07 30	174		"		"	"
2:08	180		"		"	"
2:08 30 "	180		"		"	"
2:09 30			"		"	Amyl. Alc. 1/10%
2:10	stopped		"		39.5	"
2:10 45	" "				"	Pure
2:11 30	began again but very irregular and weak.				"	"

Note: Heart and lungs covered with large ecchymoses. Measurements were taken very nearly once in 30 seconds.

RESULTS:

At 36° a fall from 193.5 C.C. to 43 C.C. or red. of 150.5 C.C.
" 37 " "   "   184.5 "   " 25 "   "   "   " 159.5 "
" 38 " "   "   182   "   " 1 "   "   "   " 181   "

This gives an average of 163.366 reduction at 37° in the amount pumped around in 30 seconds. The last measurement under Amyl. Alcohol taken at 2 o'clock 9 min. and 30 seconds, is not reliable, because as the drug produced a reduction of 181 C.C. at 38° it would certainly from what we know have had a greater effect than 180 C.C. at 39.5° Moreover the heart did not in the latter case make a recovery.

EXPERIMENT N.II, June 3rd, 1886. Amyl Alcohol 1/10%. Defibrinated calf's blood used. Dog under Morphine and Curare. Heart isolated 3.15 P. M. remained good for 31 min.

Time	Amount pumped in 30 sec.	Temp.	Kind of blood.
3:20	198 C.C. "	36°	Pure
3:20 35 sec.	199 " "	"	"
3:21 15 "	198 " "	"	"
3:21 45 "			Amyl Alc 1/10%
3:22 30 "	48.5" "	"	"
3:22 45 "	stopped "	"	"
3:23			Pure
3:23 45 "	40.5" "	36.5	"
3:24 30 "	89 " "	"	"
3:25 15 "	123 " "	"	"
3:26	142 " "	"	"
3:26 40 "	183.5" "	"	"
3:27 15 "	192 " "	"	"
3:28	193 " "	37	"
3:29 35 "		"	Amyl 1/10%
3:30 10 "	30 " "	"	"
3:30 40 "	stopped "	"	"
3:31 30 "	"		Pure
3:32 15 "	38 C.C. "	"	"

Time	Amount pumped in 30 sec.	c.m.	Kind of blood.
3:33	50 C.C. "	37.5°	Pure
3:33 45 sec.	92 " "	"	"
3:34 15 "	104.5 " "	"	"
3:35	124.5 " "	"	"
3:36 45 "	148 " "	"	"
3:37 30 "	180.5 " "	38	"
3:38 25 "	189 " "	"	"
3:40	190 " "	"	"
3:40 45 "	189 " "	"	"
3:41 30 "			Amyl Alc. 1/10%
3:42 15 "	10 " "	38	"
3:42 50 "	stopped "	"	"
3:43 30 "		38.5	Pure
3:44	20 " "	"	"
3:44 45 "	38 " "	"	"
3:45 30 "	49.5 " "	"	"
3:46 15 "	104 " "	"	"
3:47	118 " "	39	"
3:47 45 "	124 " "	"	"
3:48 30 "	148 " "	"	"
3:49 15 "	168 " "	"	"
3:50	180 " "	"	"
3:50 30 "		"	Amyl Alc. 1/10%

Time	Amount.pumped in 30 sec.	Temp.	Kind of Blood.
3:51	15 sec. 0 or 1 C.C. heart pumped around a few drops and then ceased beating, pure blood was turned on but did not recover it.	39	Amyl alcohol 1-10%

Lungs very oedematous, heart and lungs covered with large ecchymoses all over.

In this experiment 1/10 of Amyl alcohol in defibrinated calf's blood affected the isolated dog's heart in the following manner:

A fall from 198 C.C. to 48.5 C.C. or red. of 149.5 C.C. at 36°) In
" " " 193 " " 30 " " " " 163 " " 37°) 30
" " " 189 " " 10 " " " " 179 " " 38°) sec.

This gives an average reduction of 163.83 C.C. at 37° in 30 sec. judging of effect of 1/5% Amyl alcoholized blood from that of 1/10% by doubling above result one obtains the following 327.66 C.C. reduction in 30 seconds at 37°.

EXPERIMENT June 4th, 1886 Amyl alcohol 1/10% in defibrinated dog's blood. Animal under Morphine and Curare. Heart isolated 1:54 P.M. under experimentation 30 minutes.

Time	Amount pumped in 30 sec.			Temp.	Kind of blood.
2	196 c.c.		"	35.5°	Pure
2:01	197	"	"	"	"
2:02	198	"	"	36	"
2:03	198.5	"	"	"	"
2:03 30 sec.					Amyl Alc. 1/10%
2:04	59	"	"	"	"
2:04 30	"			"	Pure
2:05 30	"	112	"	"	36.5	"
2:06 30	"	174	"	"	"	"
2:08	194	"	"	"	"
2:09	198	"	"	"	"
2:10	198	"	"	37	"
2:10 15	"				Amyl Alc. 1/10
2:11	37	"	"	"	"
2:11 30	"				Pure
2:12 30	"	65	"	"	"	"
2:14	124	"	"	37.5	"
2:15	186	"	"	"	"
2:16	196.5	"	"	"	"

82

Time	Amount pumped in 30 sec.	Temp.	Kind of Blood.
2:17	197 C.C. "	38	Pure
2:17 30 sec.			Amyl Alc. 1/10%
2:18 15	" 18 " "	"	"
2:19			Pure
2:20	59 " "	"	"
2:21	140 " "	30.5	"
2:22	182.5 " "	"	"
2:23	196 " "	"	"
2:24	196.5 " "	"	"
2:24 20	"	39	Amyl Alc. 1/10%
2:25	a few drops then stopped "	"	
2:25 20			Pure
2:26	40 C.C. "	"	"
2:27	89 " "	"	"
2:29	180 " "	"	"
2:29 45	186 " "	39.5	"
2:30	187 " "	"	"
2:30 20			Amyl 1/10%
2:31	heart stopped pumping around. "		
2:31 40	Did not recover under pure blood		Pure

RESULTS of Amyl alcohol 1/10% in defibrinated dog's blood.

At 36° a fall from 198.5 C.C. to 59 C.C. or red. of 139.5 C.C.	)	
" 37° " " " 198 " " 37 " " " " 161 "	) In) amount
" 38° " " " 197 " " 18 " " " " 179 "	) pumped) around
" 39° " " " 196.5 " " 0 " " " " 196.5 "	) in 30")

The effect observed at 2:25 P.M. at 39° is counted in on making up the proceeding table of results, since the heart made a fair recovery under pure blood although it stopped beating before. It is probable that the effect of 1,10% Amyl alcohol at 39° is even greater than 196.5 C.C. reduction, a greater effect could not be determined as there was not more than 196.5 C.C. pumped around under pure blood.

Average reduction 159.83 C.C. at 37° in amount pumped around in 30 seconds.

EXPERIMENT .I June 2 d, 1886 Amyoxyhydrate 1/20% in defibrinated calf's blood. Heart isolation completed 1:30 P.M., remained good for 25 min.

Time	Amount pumped in 30 sec.	Temp.	Kind of blood.
1:35	180 C.C. "	35	Pure
1:36	184 " "	"	"
1:36 30 sec.	182 " "	35.5	"
1:37 15	182 " "	"	"
1:38	183 " "	36	"
1:38 30	" "	"	Amyl Alc. 1/20%
1:39 30	" 106 " "	"	"
1:40	80 " "	"	"
1:40 30	" 5 " "	"	"
1:41	stopped "	36.5	"
1:41 30 sec.		"	Pure
1:42 30	10 " "	"	"
1:43	72 " "	"	"
1:44	101 " "	"	"
1:45	140 " "	37	"
1:45 30	" 162 " "	"	"
1:46	180 " "	"	"
1:46 30		"	Amyl Alc. 1/20%
1:47 30	100 " "	"	"

Time	Amount pumped in 30 sec.			Tmp.	Kind of blood.
1:48		20 C.C.	"	37	Amyl Alc. 1/20%
1:48 30	stopped		"	37.5	"
1:49				"	Pure
1:49 45		35 C.C.	"	"	"
1:50 35	"	94	"	"	38	"
1:51 30		110	"	"	"	"
1:52		129	"	"	"	"
1:53 30	"	162	"	"	"	"
1:54 30	"	174	"	"	"	"
1:55			"	"	Amyl Alc. 1/20%
1:55 45	"	87	"	"	"	"
1:56 30	"	40	"	"	"	"
1:57 30	stopped		"	"	"
1:58 30			"	39	Pure
1:59 30		64	"	"	"	"
2: 30	"	98	"	"	"	"
2:02		122	"	"	"	"
2:03		160	"	"	"	"
2:04		168	"	"	"	"
2:04 30	"				Amyl Alcohol
2:05 15		74	"	"	"	"
2:06		14	"	"	"	"
2:07	Heart irregular			"	"

Time	Amount pumped in 30 sec.	C.c.m.	Kind of blood.
2:07 30 sec.	Heart irregular	39	Pure
2:08	Did not recover under pure blood.		

RESULTS:

Effect of 1/20% of Amyl alcohol in defibrinated calf's blood upon the isolated dog heart.

At 36° a fall from 183 C.C. to 106 C.C. or red. of 77 C.C.) In
" 37° " " " 180 " " 100 " " " " 80 ") amount
) pumped
" 38° " " " 174 " " 87 " " " " 87 ") around
) in 30
" 39° " " " 168 " " 74 " " " " 94 ") seconds

Average taken from the measurements at 36° 37° and 38° is 81.33 C.C. reduction at 37° in amount pumped around in 30 seconds.

Time	Amount pumped in 30 sec.			Temp.	Kind of blood.
3:25	198 C.C.	"	"	35.5°	Pure
3:26	199	"	"	"	"
3:27	200	"	"	36	"
3:28	198	"	"	"	"
3:28 30 sec.			15	"	C_2H_5O 1/5%
3:29	48	"	"	"	"
3:29 30	" stopped		"	36.5	"
3:30			30	"	Pure
3:31	98 C.C.		"	"	"
3:31 30	" 146	"	"	"	"
3:32 30	190	"	"	37	"
3:33 30	192	"	"	"	"
3:34	191	"	"	"	"
3:35			15	"	Amyl Alc. 1/5%
3:36	30	"	"	"	"
3:36 30	" stopped		"	37.5	"
3:37			30	"	Pure
3:37 30	" 90	"	"	"	"

Time	Amount pu... in 30 sec.			Temp.	Kind of blood.
3:38	136 C.C.	"		37.5	Pure
3:38 30 sec.	180	"	"	38	"
3:39	178	"	"	"	"
3:39 30	179	"	"	"	"
3:40			15	"	Amyl alc. 1/5%
3:40 30	8	"	"	"	"
3:41	stopped		"	"	"
3:41 30	"		30	38.5	Pure
3:42	54 C.C.	"	"	"	"
3:42 30	" 112	"	"	"	"
3:43	110	"	"	"	"
3:43 30	106	"	"	"	"
3:44	120	"	"	"	"
3:44 30	" 154	"	"	39	"
3:45	169	"	"	"	"
3:45 30	176	"	"	"	"
3:46	177	"	"	"	"
3:46 30	" 176	"	"	"	"
3:47			15	"	Amyl Alc. 1/5%
3:47 30	" stopped		"	"	"
3:48			"	"	Pure

Time	Amount pumped in 30 sec.	Temp.	Kind of Blood
3:49	heart began to pump around again but so weak that the experiment could not be continued.	39.5°	Pure

As it was found that 1/5% solution of Amyl Alcohol could not be left turned on and flowing through the heart for 30 seconds without causing its death. It was attempted to allow the measurements under Amyl alcohol only 15 seconds with a view to observing what its effects in that period of time would be.

The following are the results of above experiment.

At 36°C a fall from 198 C.C. to 48 C.C. or red. of 150 C.C.)After
" 37° " "   "   " 191 "   " 30 "   "   "   " 161 "   )flowing
)through
)alcohol
" 38° " "   "   " 179 "   " 8 "   "   "   " 171 "   )for 15
)seconds
" 39° " "   "   " 176 "   " 0 "   "   "   " 176 "   )

Average 160.66 C.C. reduction at 37°C. after an exposure to Amyl alcohol and counting 15 seconds.

XPERIMENT ... 25th, 1886 1/5. Amyloxyhydrate $C_?H_?O$ in defibrinated calf's blood. Isolation completed 2:15 P.M. Heart ... for experimentation for 26 minutes.

Time	Amount pumped in 30 sec.	Temp.	Kind of blood.
2:18	192 C.C. "	36° C	Pure
2:19	192 " "	"	"
2:20	192 " "	"	"
2:21	15	"	Amyl Alc. 1/5%
2:21 30 sec.	40 " "	"	"
2:22	stopped "	"	"
2:23	30	36.5	Pure
2:24	50 " "	"	"
2:25	140 " "	"	"
2:26	162 " "	37	"
2:27	184 " "	"	"
2:28	189 " "	"	"
2:29	190 " "	"	"
2:30	190 " "	"	"
2:31	15	"	Amyl Alc. 1/5%
2:32	29 " "	"	"
2:32 30	" stopped "	"	"
2:33	30	"	Pure
2:33 30	" 20 " "	37.5	"

Time	Amount pumped in 30 sec.			Temp.	Kind of blood.
2:34	58	"	"	37.5	Pure
2:34 30 sec.	85	"	"	"	"
2:36	108	"	"	"	"
2:37	142	"	"	38	"
2:38	168	"	"	"	"
2:39	182	"	"	"	"
2:40	183	"	"	"	"
2:40 30	" 182	"	"	"	"
2:41		15		"	Amyl Alc. 1/5%
2:41 30	10	"	"	"	"
2:42	stopped		"	38.5	Pure
2:43	did not recover perfectly under pure blood.			"	"
2:45	ceased beating.			"	"

RESULTS:

At 36° a fall from 192 C.C. to 40 C.C. or red. of 152 C.C.

" 37° " " " 190 " " 29 " " " " 161 "

" 38° " " " 182 " " 10 " " " " 172 "

Average reduction 161.66 C.C. at 37° in amount pumped around in 15 seconds.

1/5% of Amyl alcohol as can be seen from the above is too

ution to experiment with. This force was a solution
lowed turned on and flowing through the heart for 30
out inevitably producing its death. It was therefore
to flow through for 15 seconds, hoping that we might
f the effects that would be produced in 30 seconds by
results. A priori this is not the most correct way of
under the conditions nothing was left to do but one of
a) either reduce the time through which the alcohol
h the heart or (b) reduce the strength of the solution
see both methods led to approximately the same re-
rtially supplemented each other.

RESULTS OF EXPERIMENTS WITH AMYL ALCOHOL.

$1/5\%$ $C_5H_{12}O$. Reduction in amount pumped around by isolated leg part in 15 seconds at $37°$ effected by $1/5\%$ solution Amyl alcohol

May 28th, 1886, 161.66 C.C.)   Average   161.16 C.C.
)
June 1st,   " 160.66 ")   at $37°$ in 15 seconds.

$1/10\%$ $C_5H_{12}O$ Reductions under 1,10% Amyl alcohol.

May 27th, 1886, 159.   C.C.) Average 161.508 C.C.
)
June 3rd,(I)" 163.366   "   )   reduction in
)
"   " (II)" 163.83   "   )   30 seconds
)
"   4th, " 159.83   "   )   at $37°$

If we may judge from two experiments with $1/5\%$ $C_5H_{12}O$ and compare them with the general result of four experiments with $1/10\%$ $C_5H_{12}O$ it is evident that $1/5\%$ will cause the same reduction in 15 seconds that $1/10\%$ did in 30 seconds.

$1/20\%$ $C_5H_{12}O$; Reduction observed in one experiment in which $1/20\%$ Amyl alcohol was employed.

81.33 C.C. reduction in 30 sec. at $37°$

Average of one experiment with $1/10\%$ Amyl Alcohol which was started at a high temperature and gradually lowered.

June 4th, 1886, 158.83 C.C.

reduction at $37°$ in amount pumped around in 30 sec.

www.ingramcontent.com/pod-product-compliance
Lightning Source LLC
Chambersburg PA
CBHW022115230426
43672CB00008B/1395